DHH Xmas 1896

P9-CNH-790

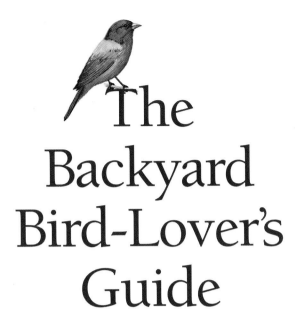

The Backyard Bird-Lover's Guide

Jan Mahnken

Color Illustrations by Jeffrey C. Domm

Foreword by René Laubach, Director, Berkshire Sanctuaries
Massachusetts Audubon Society

STOREY

A Storey Publishing Book

Storey Communications, Inc.

The mission of Storey Communications is to serve our customers by publishing practical information that encourages personal independence in harmony with the environment.

Edited by Deborah Burns
Design and production by Cynthia N. McFarland
Production assistance by Allison Cranmer
Black and white drawings by Kimberlee Knauf
Indexed by Northwind Editorial Services

Copyright © 1996 Storey Communications, Inc.
Some of the material in this book was previously published by Garden Way Publishing in *Feeding the Birds* (1983) and *Hosting the Birds* (1989), both by Jan Mahnken.

Birds pictured on front cover, clockwise from top left, ruby-throated hummingbird, song sparrow, eastern bluebird, scarlet tanager pair, northern oriole; on spine, barn swallow; on back cover, northern flicker

All rights reserved. No part of this book may be reproduced without written permission from the publisher, except by a reviewer who may quote brief passages or reproduce illustrations in a review with appropriate credits; nor may any part of this book be reproduced, stored in a retrieval system, or transmitted in any form or by any means — electronic, mechanical, photocopying, recording, or other — without written permission from the publisher.

The information in this book is true and complete to the best of our knowledge. All recommendations are made without guarantee on the part of the author or Storey Communications, Inc. The author and publisher disclaim any liability in connection with the use of this information. For additional information please contact Storey Communications, Inc., Schoolhouse Road, Pownal, Vermont 05261.

Printed in Canada by Métropole Litho

10 9 8 7 6 5 4 3 2 1

Library of Congress Cataloging-in-Publication Data

Mahnken, Jan.
      The backyard bird-lover's guide / by Jan Mahnken
         p.   cm.
      "A Storey Publishing book."
      Includes index.
      ISBN 0-88266-927-3 (pbk. : alk. paper)
      1. Bird attracting. 2. Bird feeders. 3. Birds—Feeding and feeds. 4. Birds—Nests. 5. Birds—Behavior.
      I. Title.
      QL676.5.M25   1996
      639.9'78—dc20   96-5391
                                                                          CIP

For Bud, Bob, and Tom

CONTENTS

FOREWORD

AMONG THE EARTH'S MULTITUDE of creatures, birds are surely some of the most beloved. They are revered in song and verse, sports teams are named for them, and each of our states long ago adopted one as its own. Why do birds beguile us so?

Their beauty and vocal virtuosity are undeniable. We marvel at the cardinal's brilliant red against snow-covered ground and the hermit thrush's ethereal song. Even those who do not profess to be "bird watchers" are inspired by the magnificence of a red-tailed hawk or the majesty of our national symbol, the bald eagle. Beyond the sweetness of their dawn chorus and their beauty, birds are also fascinating in habit. It is not surprising then that so many millions of people feed, watch, photograph, and otherwise enjoy these wonderful creatures.

For everyone deeply interested in birds there seems to have been one special bird or avian event that provided inspiration. For many it resulted from feeding birds. Who isn't drawn to the winter bird feeder, curious to see what wondrous living thing has been lured from the forest or field or hedgerow to our offerings?

I can still vividly recall the spark that kindled my life-long love affair with birds when I was thirteen. Sitting one sunny midday on the steps of a wooden cabin at Boy Scout camp in southeastern Michigan, I spied a small bird that I had never seen before. It hopped about on the bare earth, sporting a reddish cap and a black horizontal line across its face. It was a warm grayish color below with varying shades of rich brown above. With the help of a beginner's bird guide I identified my first chipping sparrow.

The next year I was back at camp hoping for bigger game. One bright morning from the shrubby tangles near our tents I heard a whistled three-part refrain — "drink your tea." After a few moments of stalking about I spotted the gorgeous and rather sizable form of a male rufous-sided towhee. By now I was hooked.

This morning, I was heartened to witness the return of "our" male red-breasted nuthatch, after an apparent absence of many weeks. While the blizzard of '96 piled the snow up to and over our first-story windowsills, this handsome sprite — so bright rufous below, with a shiny dark cap — made repeated sallies to the hanging tubular feeder, joined by a host of black-capped chickadees.

My work now, for the Massachusetts Audubon Society — which celebrates its Centennial this year — sometimes takes me places far from home, but wherever I go, I'm thankful for the tremendous pleasure that birds give me.

To say that birds have given many millions of people countless hours of enjoyment and even inspiration may be an understatement. If you're not already hooked, this book may help turn a casual interest into a healthy obsession. Perhaps your own spark is just moments away from alighting outside your window.

René Laubach
Sanctuary Director
Berkshire Wildlife Sanctuaries
Massachusetts Audubon Society

PART I
FEEDING AND
HOUSING THE BIRDS

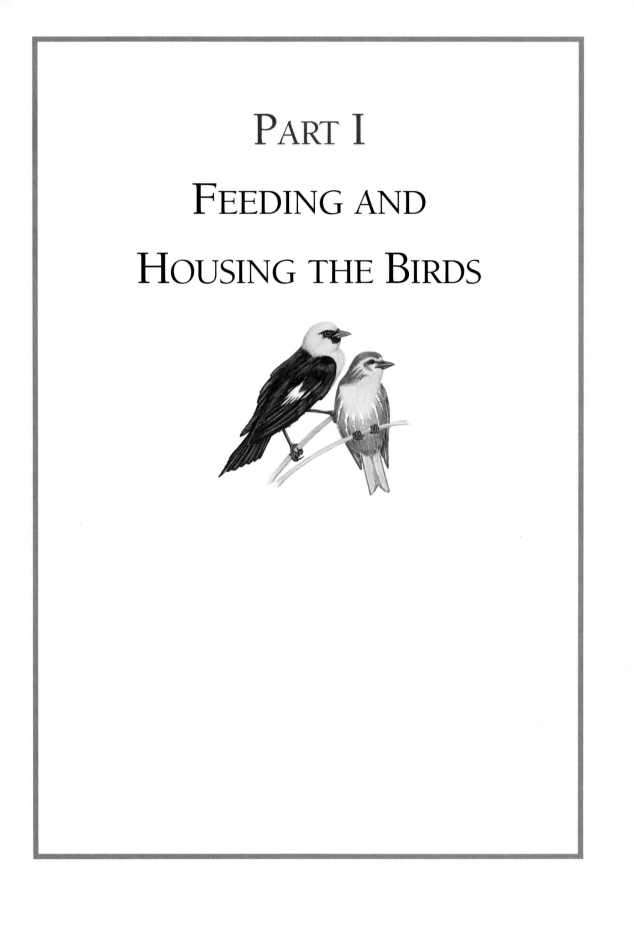

-1-

Why Feed the Birds?

Allen's hummingbird

PROBABLY THE BEST REASON to start feeding birds is that you *like* birds. You enjoy having them around for a variety of reasons, mostly aesthetic. Their songs, their squawks, their comments drifting through the open windows on a summer day keep you informed of what's going on outside whether you can see it or not. Loafing on the porch steps after dinner, we watch the swallows darting about the pasture, the hummingbirds visiting petunias, the kingbird sitting motionless on a fence post. The evening grosbeaks, the cardinals, and the blue jays brighten up a winter day when the rest of the world seems somber in black and white and shades of gray.

Wherever you live, you'll find a wide variety of wildlife, and much of that variety comes in bird form. Globally, there are over 8,500 living bird species, 950 of them on our continent alone. Many are highly visible and relatively easy to attract if we provide food, water, and shelter.

Interest in birds is a worldwide phenomenon. Each summer a friend of mine in Washington, D.C., sends me newspaper clippings reporting the progress of the peregrine falcons that nest annually on a rooftop in downtown Baltimore. On the other side of the world, a female spoonbill duck that nested for five successive years in an artificial pond across from the Imperial Palace in Tokyo attracted even more notice. Each

A Popular Pastime

According to estimates, some twenty-one million people engage in bird watching, a national pastime second in popularity only to gardening.

year she led her brood across the busy six-lane highway to spend the summer in the palace's moat. When Tokyo police thought her crossing was imminent, they posted warnings, and the actual event literally stopped traffic, as well as making the newspapers and network television.

Once embarked on the project of feeding the birds, you're likely to become more and more interested in them. Observing their behavior is easier because you've lured them into range. The beauty of the whole process is that it can be indulged in at whatever level appeals to you. It's a hobby you can begin with a minimum of fuss at any time of the year and continue at whatever rate you choose.

American robin

GETTING STARTED

How do you begin? That's the easiest part. You don't actually even need a feeder. Suet broken into small pieces attracts birds. From that point on, it's a matter of refinement. Add a feeder. Provide water in a shallow, rough-textured container. Buy some bird seed, and you're on your way.

Be Consistent

It's as easy as that? Well, almost. Once you start, especially in lean times for the birds, there's the moral obligation to keep going, to be consistent. If birds learn to depend on your handouts, you ought to make sure that they continue to eat even if you're away from home. That can be arranged either through careful selection of food and feeders or through having someone else stock the feeders in your absence. Some people plan a feeding program that runs year-round; others supply food only when the natural supply is inadequate.

The Economics of Feeding Birds

Don't forget that it's an expense — and even if you consider having an inordinate number of dependents a charity, the IRS doesn't. It must be admitted that there's another economic problem in attracting birds. They sometimes feed on crops intended for human or livestock consumption. We've planted a number of dwarf fruit trees in the side yard, and last year the robins consumed two-thirds of the cherry crop. (I gave the other cherry to my son Tom.) Catbirds take up summer residence in our raspberry patch, but fortunately there's enough produce to satisfy all of us. That seems to be the most effective way of dealing with bird depredations, at least on an individual level.

PRACTICAL BENEFITS

If your conscience demands practical reasons for attracting birds, consider their usefulness in reducing insect populations. Those starlings parading around the lawn will eat Japanese beetle larvae. Not only that, their beaks probing the ground for breakfast pick up and spread spores of a disease that helps control the beetles. The often maligned starling also goes after gypsy moth larvae, a delicacy scorned by most of the native birds. A northern (Baltimore) oriole consumes seventeen hairy caterpillars a minute, and a pair of flickers will polish off five thousand ants as a first course. Even birds that are primarily seed eaters feed their nestlings a diet high in insects.

Their appetite for fresh fruit in season may compete with ours, but we're surely willing to allow the birds to harvest the berries of poison ivy and sumac and the seeds of weeds. If it weren't for the American goldfinch, we'd be up to our

The American goldfinch loves thistles.

Feeding birds is both entertaining and educational.

hips in thistles and catnip. There's another angle to consider. By various means, birds do a fine job of planting trees, shrubs, and other kinds of vegetation. People often find rather remarkable plants growing under their bird feeders. Birds have done their part in replacing the hedgerows that disappeared as a result of our passion for clean cultivation. The jay who buries an acorn and forgets where he cached it has helped to propagate oaks. Thus to a certain extent, the birds help maintain their own food supply and that of other animals. At the same time they reduce soil erosion.

Fun for All Ages

You may be able to rope children into the project. Our boys have shared in building feeders and birdhouses. They've helped keep them filled. They've scrambled up ladders to replace nestlings and decorated improvised hospitals with greenery in an effort to keep the patients comfortable. A black swan at a public park instilled respect in my son Bob by a well-placed peck, and an incapacitated hummingbird taught Tom that transistors aren't the only astonishingly complex items to be found in small packages. Firsthand experience sensitizes a child to the needs and demands of living creatures as watching cartoons never could.

Because feeding birds bring them near enough to observe at close range, it can add another dimension to the lives of people whose horizons are unavoidably limited. Radio and television have been boons to people confined indoors because of age, illness, or handicaps, but they can't hold a candle to the pleasures of observing live creatures right outside the window. I'll never forget how much my maternal grandmother enjoyed watching the birds that came to the feeders on her front porch. Their antics occupied her during long hours as she recuperated from a broken hip.

Helping the Environment

The clinching argument for feeding birds? I think we have a responsibility to them. We've interfered extensively with their environment as our population has expanded. It's quite true that the extinction of species occurs naturally; man's interference, however, has accelerated the rates alarmingly. We owe them. This debt can be paid in part by making our property more hospitable to birds.

-2-

The Initial Plunge

I'VE MENTIONED that you can start a bird-feeding program at any time of the year, and that's perfectly true. It's also true, at least in some parts of the country, that you'll get off to a more auspicious start in some seasons than in others.

In the middle tier of states, north to the latitude in which Massachusetts is the easternmost state, bird feeders get the least business from the summer nesting season until late in October. Natural food supplies are abundant at that time, and by August migratory birds are beginning to get into the mood for setting out on the road and are restless. If the notion of catering to the birds strikes you during that slack period, October 1 is a good arbitrary date for beginning. Start a month earlier in the northernmost tier of states. The presence of even a few birds at that time may direct transient visitors to your feeder during migration and signal winter residents as well. Because there isn't much turnover in resident birds in the deep South, Texas, and the Southwest, timing is less critical in those areas.

A peregrine falcon dives after its prey.

Other periods of heavy feeder use occur during the winter months after a sudden drop in temperature or during snowstorms and periods of heavy snowcover.

ATTRACTING BIRDS

How will the birds know you've elected to become their patron? It's often the scavengers — gulls, pigeons, starlings, grackles, house sparrows — who are first at the scene of a brand-new food supply. Who discovers your bird feeder first will depend on the season and your location.

For more on what specific birds like to eat, see the chart "Who Eats What," starting on page 288.

Chickadee, chipping sparrow, and titmouse appreciate good cover.

If you have trees nearby, the jays and chickadees will be quick to discover your feeder, too. Cover is an important aspect of attracting birds to your premises. Indeed, I find one of the more engaging characteristics of birds is their scorn for absolutely immaculate yards. They consider your brush pile, the briar patch, and the tangle of weeds behind the garage totally irresistible. If you can provide some trees and shrubbery, so much the better. Birds are an ornament more common to a yard that is less than spic and span.

WHAT, HOW, AND WHERE TO FEED

You've decided to begin. You want birds in your yard, and the average city or suburban lot simply doesn't provide enough natural food to support many birds. Birds have enormous appetites relative to their size. East of the Mississippi River, for example, an acre of land will support only one to ten birds; the average is four. It stands to reason, given the figures, that there's not going to be much bird activity in areas of dense human population without the help of those humans.

You just happen to have some leftover bread. You've eliminated the driveway and the walks as possible sites for the banquet since vehicles and pedestrians make those areas inappropriate except for occasional gestures. Where to serve the feast? What are the possibilities?

Don't Stop Too Soon

We more often think about feeding birds in cold weather when their needs are more obvious. Fuel keeps them warm, and food provides that fuel. They're less likely to freeze to death if there's a certain amount of shelter from the weather (evergreens are especially useful to them) and plentiful food. But if you elect to stop with the return of mild weather, make sure you aren't premature. A couple of weeks after the usual time of the last snowstorms or last average frost is a good target date.

Remember that most of the birds attracted to feeders are primarily seed and/or insect eaters. Birds, whose eyes are proportionately larger than the eyes of most mammals, have superb vision. Experimental studies have compiled considerable evidence to indicate that they are attracted to white foods such as suet or white bread. These foods are therefore effective in introducing a new feeding area to birds. Although you may attract them originally with crumbs of white bread, this food is not healthy for birds. To keep them happy and healthy, you'll need to feed them something more substantial.

Fox sparrow

Seeds are available in blended mixtures or separately. Depending on the components of the mix, five pounds of blended seed will be less expensive than sunflower seeds. Beef suet substitutes for insects and can be fed straight from the butcher shop or in cakes made from rendered suet, with seeds added. You can whip these up yourself in the kitchen or buy them ready-made where you get your seed.

We'll go into more detail about specific foods for specific birds later on. Do remember that wet feed can become moldy and infect birds that eat it with the deadly fungus *Aspergillus fumigatus*. Clean feeders regularly. It's recommended that they be scrubbed with a bleach solution and sprayed with a product such as Lysol. I confess that I know no one who does that. If uneaten food is removed, perhaps fresh air and sunshine do the trick of sanitizing. If you do use a disinfectant, let the feeder air-dry thoroughly before you fill it with seeds.

Supplied with your bird groceries, your next step is to dispense them. Before rushing out to buy the avian equivalent of a five-star restaurant, you might consider a few alternatives. Look around your residence. The flat roof of our side porch is accessible from the guest room windows and provides a wide flat expanse on which birds could feed comfortably, and we could see them easily. It struck me as a great possibility for bird feeding. There was just one hitch. The roof was the favorite sunbathing area for our cats, at all seasons. If you have such an area and it isn't accessible to felines, it might be just the ticket.

PREDATORS AND ROBBERS

One of your primary responsibilities when feeding birds is safety. You have to be sure you don't place bird feeders within easy reach of predators or of mischievous rascals like squirrels.

Cats

Cats are the first predators that come to mind when we talk of songbirds in the garden. Few people are indifferent to cats. You like 'em or you don't. I happen to belong in the first category, and my attitudes reflect that. Despite what their detractors say, cats are not evil. They are simply cats. It's as pointless to blame them for stalking birds as it is to praise them for catching mice or rats or other four-footed beasts you want to eliminate.

Dedicated cat lovers though we are, we don't want our pets, or the pets of our neighbors, killing the songbirds. The birds themselves take certain precautions. So do we. They seem highly effective.

Keeping Nesting Boxes Safe

Cats are a problem around nesting boxes; they sometimes lie in wait and leap up to grab a bird as it flies from the entrance hole. Mowing the grass in a twelve-foot circle around the nesting box will prevent cats from slinking up to surprise their prey.

Precautions

For a start, our free-standing feeders are mounted on one-inch galvanized pipe driven into the ground. We adopted this tactic because a couple of our cats took to having their mid-day siestas *in* the feeder trays when the feeders were mounted on stout wooden posts.

Our suspended feeders are carefully placed so that cats can't ambush the birds. We use branches only if cats can't get above them. Ground feeding is conducted only in open areas, and the water is out in the open, too, so the cats can't lie

The neighborhood cats may be songbirds' major predators.

Tips to Deter Cats and Squirrels

- ◆ Keep feeders at least eight feet from trees, fences, or buildings from which squirrels can jump to their dinner
- ◆ Free-standing feeders on pipe will discourage cats from climbing
- ◆ Grease wooden posts or sheathe them in metal
- ◆ Mount metal cones underneath feeders to prevent creatures from climbing up
- ◆ With a trolley feeder on a clothesline, put lengths of derelict plastic hose over the rope on either side of the feeder so the squirrel's feet have no purchase
- ◆ If you enjoy feeding squirrels, try putting their food — corn or whatever — in a separate area

hiding nearby. There are suitable trees and shrubs close enough where the birds can preen.

Whenever we notice fledglings just out of the nest, we bring the cats *and* dogs indoors because fledglings are easy prey. If the parent birds are in evidence, you can bet the cats won't be. I've seen our pets racing for cover, pursued by irate barn swallows, the cats loudly protesting their innocence all the time they're retreating. The birds, by the way, won't bother a sleeping cat.

Squirrels

Probably the most common intruders at feeding stations are squirrels. Although they annoy many bird lovers, squirrels are after all natives, unlike cats. The ingenuity invested in foiling squirrels has been truly prodigious. Even if you don't object to supporting them, their presence at a feeder prevents the birds from eating, often for dangerously long periods. Squirrels have become problems at urban and suburban feeders because they adapt more easily to densely populated areas than do their enemies, so they've increased their numbers greatly.

The devices available to thwart their attempts to use bird feeding stations are numerous. If you buy a feeder for birds, it's relatively easy to get a "squirrel-proof" one. They will be discussed in more detail in the following chapter. For homemade or non-squirrel-proof feeders, a few commonsense precautions will keep the situation under control (see box).

Chipmunks, rats, mice, and raccoons may also be attracted to your feeding station, but they're much more easily

Neuter Your Cat

It's both sensible and humane to neuter cats not needed for breeding purposes. It will help prevent the high incidence of strays and feral cats. If your locale is plagued with feral cats, as opposed to well-fed domestic cats, notify your game warden.

Squirrels

Songbirds are afraid of squirrels and keep their distance if the squirrels have access to feeders. In urban areas that lack hawks, owls, and human hunters, squirrels can be a colossal nuisance to people trying to feed birds. Hence the emphasis on deterring them.

discouraged than squirrels because they're not as agile. Action you take to prevent squirrels from gobbling all the goodies will prevent these other animals from becoming much of a nuisance.

Predator Birds

Cats aren't the only predators of birds. Birds themselves prey on other birds. At one time hawks and other birds of prey were shot on sight. Fortunately, a more reasonable attitude now prevails. Recently we observed a merlin strike a robin right outside our kitchen window, on a broad expanse of lawn. Yes, we were saddened at the loss of the robin. But I confess to a keen interest in the event, too. We can — and should — protect birds from danger that *we* introduce into their environment, but let's not go monkeying around with natural enemies. Hummingbirds sometimes get caught in spider webs when they're collecting nesting materials. They're occasionally swallowed by bass or bullfrogs while drinking. One birder reported seeing a hummingbird fall victim to a praying mantis! There's no need to interfere. It's all part of the food chain. Without checks on their numbers, our songbirds would run out of their food supplies.

Sentinel Birds

Birds themselves have their own methods of coping with predators. The often-maligned jays act as sentinels, calling the attention of other birds to all manner of intruders — hawks and shrikes as well as cats and other mammals. Jays and other birds use a variety of alarm calls, depending on whether the enemy is approaching by air or by foot. Open-country birds tend to be more "flighty" than woodland birds because their environment offers them less protection. Evidence suggests that brightly colored birds are less palatable than those with dull plumage. The latter often "freeze" when alarmed. Birds can often be seen ganging up on, or "mobbing," their enemies, a tactic that frequently drives the enemy away. Usually such attacks are all bluff, but birds have been known to kill snakes. Several different species may cooperate in mobbing.

SIMPLE FEEDERS

A metal cone will defeat the hungriest squirrel.

Margaret Millar, in her delightful book, *The Birds and Beasts Were There* (New York: Random House, 1967), started her feeding program on a wide ledge outside the living room windows. Such simple beginnings have much to recommend them, especially to those of us who are indolent and/or prone to procrastination. They are easily stocked and they bring the birds close enough to be enjoyed. Their limitations, unfortunately, are pronounced. To begin with, many houses don't have such convenient architectural attributes. There's also the problem of untidiness, which involves not only aesthetic but also sanitary considerations. The mess has to be cleaned up periodically. And make no mistake, birds are terrifically messy. (Ever see the havoc a budgie — *one* budgie — can create in a room?) The greater your success in attracting visitors, the more frequent and arduous your housekeeping chores.

With any feeding area adjacent to windows, you must also consider the problem of birds smashing into the glass. Owl or other predator silhouettes, sold for the purpose of preventing such calamities, don't discourage birds in time, so you've got to think of some other solution. Try tacking red ribbons from the top sills of your windows — if you don't mind how this looks. Screens remaining in place outside the window may work in some cases; sheet curtains that prevent the birds from considering the window a flying area may be even better. They have the extra advantage of masking *you*, giving you a better chance of observing your guests close up.

In addition, even if cats can't find their way to the area, rats, mice, squirrels, raccoons, and snakes (which feed on the mice!) just may. Cats and squirrels are problems enough; most of us aren't prepared to support the entire wild population of our neck of the woods.

Bird Table

A near relative of the ledge-roof approach, the bird table has many of its virtues as well as its disadvantages. It need

A bird table can be a convenient way to feed birds.

not be designed for the purpose, though it may be; any table impervious to weather will serve. It will accommodate significant numbers of birds and, since tables are portable, its location can minimize some of the problems associated with feeding the creatures actually *on* the house. Because it may be placed at some distance from the house, shyer species may be attracted, and you're less likely to have birds banging into windows. Keeping it tidy is usually easier than policing ledges and roofs but remains a necessary task.

Make It Yourself

Before rushing out to buy a feeder, take stock of the basement or barn or shed or garage. There must be *something* there with which to fashion a feeder. Even the constitutionally inept (our category) can produce a satisfactory one with a minimum of effort and expense.

If you have the time, interest, and requisite carpentry skills, you can read many books that give detailed instructions for making handsome feeders, from garden variety to sumptuously sophisticated. See page 302 for a few suggestions.

Coat the outside of the feeder with a natural stain (but remember to leave the interior bare). Birds prefer the dullness of stain to the brightness of paint; a natural finish is less threatening to them. Ideally, stain the feeder in the spring and let it weather outside for a few months before you start feeding the birds.

There is considerable evidence to suggest that a simple open tray-type feeder, similar to the bird table but mounted about five feet from ground level, may be the most effective way to

---

Easiest Feeders to Make

The simplest feeder I've seen described is nothing more than a board, preferably with a rim (the better to keep the feed from blowing or sliding off), attached to a windowsill. Dimensions, of course, are determined by the size of the window chosen.

Another simple feeder is made from a coconut. Remove the milk and saw the coconut lengthwise in half. Leave the meat inside as a treat, if you like. Drill three holes through which to thread wire or stout twine near the sawed edge so that it can be hung like a hanging pot. Fill with seed, bacon grease, and peanut butter or other delicacies, and voilà, an instant feeder, ready to go. Two instant feeders, even.

---

attract birds. Since the food will be unprotected from snow and rain, the tray should have drainage holes drilled in it. Whatever food drops through these holes will be eaten by birds that prefer to feed on the ground. Absence of a roof makes the food more visible to passing birds, and it prevents the more cautious of them, who may at first be suspicious of closed-in places, from being spooked. If birds seem reluctant to feed close to the house, the feeder can first be placed at some distance and gradually moved closer.

Steller's jay

If feeding from windowsills is your eventual goal, a trolley feeder is especially useful. Every couple of days you can inch it a bit closer to the house. The food will remain in approximately the same location, so the birds will have no difficulty in finding it.

Some stores offer inexpensive, easily assembled feeder kits. Youngsters just beginning to be interested in birds and/ or building may be pleased with such a project. With the kit they can put together a good-looking feeder with few problems and minimal frustration.

Another type of homemade feeder, dear to the hearts of den mothers and scoutmasters, is the length of log with holes drilled in it. Usually the bark is left on, the better for bird feet to maintain a grip and also, I suspect, to make it look suitably rustic and unobtrusive. Most of these logs have a screw eye attached to one end so they can be hung from a branch. Fancier models provide perches at the drilled holes. Some versions are so complicated to make that they would delight a dedicated woodworker.

This feeder doubles as a suet dispenser and as a grease-concoction dispenser. You can mix peanut butter with partially hardened bacon grease and perhaps some birdseed or cornmeal and push the mixture into the holes. This is easier said than done. Please note that this is, of necessity, a cold-weather procedure. In my experience, this particular feeder needs refilling by the time you get indoors, remove your boots, and hang up your coat. Children, however, seem to love them — at least, they love to make them. When requested to fill them, they remember urgent business elsewhere or tell you they have too much homework.

Ground Feeding

Don't neglect the ground itself as a feeding area. Some birds are reluctant to feed anywhere else. Spills from the other feeding equipment you use often provide enough food to satisfy ground feeders. If that approach seems inadequate for your particular crowd, choose an area with cover nearby. In

Ground feeders are safer where there's nowhere for a predator to hide.

soggy or snowy weather, using a canvas tarpaulin or a piece of plywood as a base might help minimize waste of food. Surfaces such as these can — and should — be cleaned regularly to maintain hygienic conditions. So few of us are prepared to go to these lengths that the birds who prefer to eat on the ground are more often left to their own devices in scrounging what they can from feeder overflow.

Seasonal Fluctuation

The population at the feeder will vary somewhat from season to season, and actual numbers may not remain constant. Many people insist, however, that they fill their feeders on a regular schedule, and find little or no difference in the amount they use at different seasons. Perhaps there is less dependence from a larger population of birds in times of natural abundance and greater dependence from a smaller population in times of natural scarcity. Or maybe it's that more people stock stations in cold weather and the birds distribute themselves over a wider area and visit multiple feeders.

PLACING YOUR FEEDERS

Okay. You've got the basics. Now you have to decide where to put them. There are several considerations to keep in mind. You want to be able to see the birds from indoors. The feeders should be in a protected location and near cover for the birds. On the other hand, you want them to be easy to reach for stocking. In areas of heavy snowfall, this can be a real problem. That feeder so conveniently located beside the driveway or front walk suddenly has a high wall of snow separating you and it after the plow or the snowthrower gets finished clearing. If it's too close to areas that are plowed, it may actually be damaged by the snow-moving equipment. Moreover, those piles of snow may make the feeders suddenly accessible to animals you don't intend to feed, or to those that have designs on the birds.

Try to ensure the feeder can be reached for cleaning and filling without the use of a stepladder. There's nothing quite like a hassle of that nature to dampen your enthusiasm for the whole business, especially in inclement weather.

-3-

Bird Feeders and Food

TWO DIFFERENT SETS of circumstances may prompt you to expand your feeding program. Perhaps you've had such success in attracting birds that they're all but standing in line to partake of the banquet. On the other hand, maybe you haven't been able to attract some of the birds you especially want in your yard. In the first case you simply need more feeders; in the second, you may need different kinds of feeders or a different menu.

Overcrowding at feeders leads to all sorts of difficulties. Some birds will fight among themselves, while many others — starlings, for example — will drive the less aggressive of their own and other species away. It's hardly a surprising phenomenon and is common in other animals, including man. Too many equals stress.

You've heard of "pecking order" in chickens. The boss chicken pecks everyone below her, and all the other chickens have a specified place in the scheme, dominant over the chickens down the line, subordinate to those above. Wild birds don't have conditions as stable as those of domestic chickens, so you'll find among them a "peck dominance" instead, depending on the situation at the time. There will always be a certain amount of friction, but it can be minimized by having a variety of foods and feeders available to your guests. A single species or individual can't then dominate so easily.

Different Preferences

Understanding birds' preferences will help minimize "undesirables" like starlings and jays at your bird feeders.

Birds that prefer ground feeding include juncos, sparrows, doves, quail, cardinals, jays, starlings, towhees, and thrashers. Some of them will fly to a platform feeder to get food but take it away to eat it. Swinging feeders discourage jays and house sparrows but will attract cardinals, chickadees, finches, nuthatches, and titmice.

See also "Who Eats What" on page 288.

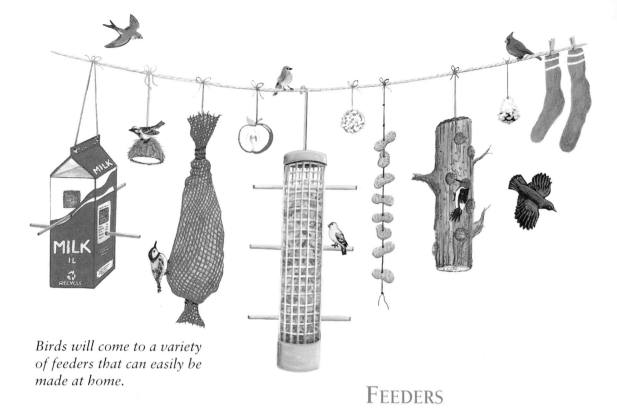

Birds will come to a variety of feeders that can easily be made at home.

FEEDERS

Let's examine some of the ready-built feeder types available. Do-it-yourselfers can build similar feeders.

Tubular Feeders

The best feeders conserve feed, are available in all kinds of weather, and are safe from disturbances, including predators. These are some of the reasons that tubular feeders have become very popular in recent years. Several manufacturers supply complete lines, ranging in price depending on capacity, amenities, and the sophistication of the feeders. Some use only metal fittings, and their higher starting prices reflect this. Some feeders must be hung; others have a pipe thread built into the bottom for post mounting. The tube itself is constructed of clear plastic and has perches at the feeding stations. It's easy to see when refilling is in order, and the feed is protected from rain or snow.

Tubular feeders are designed to hold and dispense either thistle or sunflower seeds. The latter kind will also handle cracked corn or a mix of wild bird seeds. The least expensive ones are hanging models 16 inches long and 2 inches in diameter with two perches and feeding ports near the bottom of the tube. At the same price you can buy 8-inch prefilled models.

Homemade Is Fine

It's clear that you can spend remarkable amounts of money on bird feeders, if you choose to do so. It may be fun, but it isn't actually necessary. Homemade ones, whether aesthetically pleasing or merely serviceable, will be acceptable to the birds. They have less exacting standards than we do in such matters.

In the midprice range you'll find tubular feeders with from four to eight feeding stations, each with a wooden perch. You may want brackets from which to hang the tubes. They're made of clear acrylic, are about ten inches long, and support up to fifteen pounds. They mount with screws to post, tree, or window. Some are equipped with four perches for birds waiting their turn at the feeder. Additionally, you can lock a "high impact polystyrene" seed tray, in a metallic finish, to the bottom of one line of tubular feeders to accommodate still more eager eaters.

Also useful is a 16-inch diameter squirrel guard with a "tilt-top" baffle made of clear plastic, with hardware for suspending a feeder. Some of the top-of-the-line tubular feeders are advertised as squirrel-proof, anyway. They have a metal cap and base, and some have metal perches with insulated perch guards. (Squirrels may eat wooden and plastic perches.)

Several models of tubular feeders have metal baffles within and metal reinforcing around the feeding ports themselves to prevent waste and spilling. Some of the feeders in this medium price range will hold four pounds of thistle seed, which is a lot of thistle seed. But if you want a feeder that is large (24 inches) and also squirrel-proof, be prepared for another price hike. The most complicated one I've seen will dispense sunflower, thistle, and mixed seed independently and simultaneously. It has a seven- to nine-pound capacity, nine feeding stations, a seed tray, and a squirrel guard. That's deluxe. It costs a bundle.

Feeder Posts

You like the idea of mounting a tubular feeder on a post? Such a post is available. It will fit any flat-bottomed wooden post feeders and, with adaptors, some of the tubular feeders, as well. It comes equipped with a movable plastic sleeve at the top that dumps squirrels to the ground and automatically returns to the top of the pole.

Other Popular Feeders

Besides the tubular feeders, there are others designed to be hung. One, made of nylon netting, is used to dispense thistle. Called a "thistle stocking," it is 12 inches long and holds twelve ounces of thistle. The birds love it, but remember that it doesn't protect from weather the way the tubular feeders do. The "e-z fill" is an 8-inch clear plastic cylinder, filled by rotating the roof, which rests on a 6-inch feeding tray. Next level up you'll find the "beehive" type, made of polyethylene and designed to hold sunflower seeds for the small birds that

Storage of Feed

The main requirement is to keep those feeders filled once the birds discover them. Handy quart or gallon seed dispensers with tapered spouts are available, although we make do with a coffee can. (You can squeeze one side of the rim into a V-shape to make pouring easier.) Or you can get a seed bucket, with carrying handle, that will store fifteen to twenty-five pounds of seed, depending on what kind it is. We store ours, usually in the sacks it's packaged in, in garbage cans in the barn. Unless you store bird feed in your house, you'll discover it's a good idea to use vermin-proof metal containers.

will cling to a feeder, such as chickadees, nuthatches, and finches.

Small Birds Special

Also for the small birds is the "satellite" feeder, 6 inches in diameter, advertised as spillproof, leakproof, and big-bird proof. Or you can buy a feeder with frost-free wire fencing that permits only small birds to enter. Both can be suspended or post mounted. Also available is a feeder bowl equipped with the "tilt-top" squirrel guard previously described.

Something for Everyone

The variety of feeders available is surprising. Want to cater to cardinals and other birds of similar size? There's a special feeder for them, to be mounted or hung. A hanging feeder for thistle has a cone-shaped roof and "feed-saving slots." A-frame feeders are hung through the peak of the roof to make them squirrel-proof. Tray-type feeders, to be suspended or post mounted, have a central hopper of clear plastic to dispense the seed. Depending on its size, you might be able to stock it with a supply lasting several days. A few hopper types are available in kit form.

Some hanging feeders are equipped with weathervanes and have only one open side. They turn with the wind, protecting both birds and feed. Try mounting a suspended feeder on a trolley instead of hanging it from a fixed point. It makes for ease of stocking in inclement weather, and it's a splendid way to lure the birds gradually closer and closer to viewers.

Windowpane Feeders

Windowpane feeders, made mostly or entirely of clear acrylic, are prized even above trolley feeders for their ease of filling and for bringing birds really close to observers. Most are held to the window by transparent suction cups, which lift free for filling the feeder. They range from 3½ inches to over 20 inches in size. One windowpane feeder is made specifically for dispensing thistle; another is designed to hold suet.

Suet Holders

Ways of offering suet are nearly as varied as methods of offering other bird foods. Some feeders have a special section at one end, a kind of basket arrangement, to hold it. Probably the simplest way to proceed is to put a chunk in a mesh onion bag and hang it up, suspended from a tree or the branch

Coots and Gallinules

You can branch out to the gallinules and coots, order *Gruiformes*, family *Rallidae*. Coots become quite bold about asking for bread and grain. You think jays and grosbeaks are voracious? Wait until you see how the water birds can eat. The only birds I've encountered that can top them in terms of sheer volume consumed are the emus and ostriches that gather around the cars going through safari-type game parks.

A windowpane feeder is especially delightful for children.

of a sturdy shrub or some protruding part of the house, garage, or whatever. The fact that it swings will, however, discourage some of the visitors you may want to attract. An alternative is to nail it fast to a convenient surface, preferably a tree.

Still relatively easy but more permanent (and less unsightly) a procedure is to fashion a container of heavy gauge wire. Some people suggest coating the wire with paraffin so birds' feet don't stick to the wire in freezing weather. You have the option of hanging it or attaching it to a surface.

Should you not have on hand the materials to make a suet holder, you can buy one ready-made easily and nearly as inexpensively as buying the materials. Some are designed to be mounted on a tree or the side of a building, some to be suspended. Most are cages of plastic-coated wire in various shapes and sizes. The more expensive ones have roofs to offer weather protection. Suet gets a lot of attention because it's considered a choice food by at least forty-five species of garden birds. Breads or fruits can be stashed in the holders as well as, or in addition to, suet.

HUMMINGBIRD FEEDERS

The greatest amount of attention given to the feeding of a single species is that lavished on hummingbirds. It all started with simple tubes of sugar water, a method originated by an enterprising amateur. Strips of red and orange ribbon attached to the tubes attracted the hummingbirds' attention. At some

Made to Order

If you think of a particular bird-feeding problem, chances are some manufacturer has come up with a solution for it. One feeder is made especially for bluebirds, with clear plastic sides and 1½-inch holes to exclude robins, mockingbirds, and other larger birds who might be tempted by the raisins and currants used to coax the bluebirds inside. It has a hinged roof for easy filling. Add sunflower seeds or nutmeats and the chickadees, titmice, and nut–hatches will enter, too. In the south, you'll find quail feeders. Since they're placed on the ground, they have the disadvantage of attracting rabbits, squirrels, mice, rats, and snakes.

Rufous hummingbird

Birding Bonus

Sometimes other birds learn to use hummingbird feeders. Orioles have been observed refreshing themselves at the devices.

point, a birder trying to attract hummingbirds to a vial of sugar water attached a flower to its neck. The hummingbirds were astonished at the quantities of nectar in that wondrous flower. Many ready-built feeders now come equipped with plastic flowers.

Catering to hummingbirds is strictly a seasonal occupation except at our southernmost borders. You can start out with a two-ounce feeder, a bee guard, and nectar. From there on, the sky's the limit. There's one "Nectar Bar" that attaches to a window. You can feed anywhere from one to eight birds with a single feeder holding from two ounces all the way up to a whopping sixty-four ounces. Some of them have canopies to shelter the birds from rain. The price range goes up, naturally. You can add a sturdy hanging chain with a clamp that makes it easier to detach any hanging feeders for refilling. *See* page 27 for what to feed hummers.

A friend of mine in New Mexico, Donna Sierra, told me about a visit she made to friends living near the Gila River who had hung out an ordinary hummingbird feeder. It had at least thirty hummingbirds around it all day long, and they had to replenish it three or four times a day.

BIRD HABITS

Some birds will be casualties in bad weather despite your best efforts. Either they don't come to the feeder at all, or there's not enough shelter. (We'll get to shelter you can provide for them in following chapters.) This thinning of the ranks is regarded by some ornithologists as nature's way of checking bird populations when a series of mild winters has prompted birds to overextend their normal ranges.

You'll observe what we consider thievery in birds at all seasons. Some birds are more apt to engage in the activity than others, perhaps because they're more competent in its pursuit. They raid caches made by other birds. Jays often carry food away and hide it, apparently for use in leaner times. They have been observed burying sunflower seeds, in neat rows, about a foot apart. The seeds sprouted. Now, if birds could only be taught to hoe. . . .

Hoarded foods are sometimes taken by watchful competitors. Among those who'll steal food stored by other species are chickadees, house sparrows, grackles, starlings, brown thrashers, and sparrows. Some of the group will steal from each other, too, often almost from the very beak of a feeding bird. House sparrows dart in and pluck food from in front of grackles. Considering that grackles frequently kill them

Brown thrasher

An injured raptor may show up at your feeder, which may enable it to survive the winter.

— and sometimes other birds as well — with a swift blow to the head, this is a decidedly risky business. They never seem to learn that grackles have short fuses. The grackles, after dispatching their victims, either carry them off or devour their brains on the spot. Another distressing behavior, to be observed in times of really severe food shortages, is cannibalism among songbirds. Keep those feeders stocked!

Sometimes injured birds, including hawks, will visit feeders. At migration time, such birds may be unable to undertake a long journey. It's possible that your feeder will enable them to survive the winter. Some birds sometimes molest, sometimes shun injured or oddly colored birds (albinos, for example, are not uncommon). Most of them tend to ignore injured, hungry hawks at feeders, but grackles have been observed to heckle them.

New Foods

Birds tend to be suspicious of new foods, but if they're accustomed to eating in a particular place, they're likely to test an unfamiliar tidbit left there.

THE MENU

Anyone who goes to the trouble of erecting multiple feeding stations for the birds is apt to decide to try different kinds of foods. It's important to shop around for the best prices you can find because they vary widely. If you live in an area handy to a livestock feed store, you'll discover that prices are almost invariably cheaper there than at "bird departments" in supermarkets or garden centers. This is partly the result of buying in bulk, but there will even be variations from one feed store to another. My own source of supply is 30 to 100 percent cheaper on food items than is my favorite garden store, and 15 and 30 percent cheaper than some other feed stores.

Thistle

I've mentioned thistle feeders, so we'll begin there. Thistle is expensive, especially if bought in small quantities. The good news is that it's so lightweight, you get a lot of seed per pound, and most people who feed a variety of foods use only a few pounds of thistle a year. It goes a long way if it's fed from feeders designed for it.

Our friends the Wickwires, just up the road, stock five or six feeders with cracked corn, mixed wild birdseed, sunflower seed, and thistle year-round for a large number of birds. Though they use fifty pounds of sunflower seed in a little over a month (and correspondingly large amounts of the corn and mix), they tell me that ten pounds of thistle lasts them about a year. Their thistle feeder is kept full at all times; no rationing.

A certain economy is possible by buying in bulk, and you might want to club together with other feeders of birds on costly items.

Among the birds that dote on thistle are finches, sparrows, chickadees, titmice, towhees, and juncos. Mourning doves like it, but they can't negotiate thistle feeders. Considering their appetites, it's just as well.

Thistle

Thistle is more properly called niger, which is a member of the thistle family, and is often sold under that name. The seed is imported from Africa, primarily Ethiopia. One of its advantages, mitigating its cost, is that it's impervious to wet weather.

Sunflower Seeds

Sunflower seed is another food often fed separately. It's popular with so many birds that it's not only more economical but more convenient to buy it in large quantities. Sunflower kernels (already shelled) are less messy, perhaps, but close to prohibitive in price. Most people I've talked to who feed sunflower seeds separately use fifty pounds in six weeks, year-round. That is a considerable investment, and you may decide that if the birds want sunflower kernels, they can jolly well crack the shells themselves. The many who are willing to do so include titmice, grosbeaks, chickadees, nuthatches, towhees, jays, cardinals, and blackbirds. Starlings and sparrows, which have difficulty cracking the shells, are likely to pounce on bits and pieces left by birds with stronger beaks.

Evening grosbeaks can overwhelm a feeder.

If a flock of grosbeaks descends on your feeder, the sunflower seeds will disappear in short order. For one thing, grosbeaks usually arrive in flocks, not singly, and their appetites are prodigious. One observer reported seeing a single evening grosbeak consume ninety-six (he counted) sunflower seeds in five minutes. Better buy them in fifty-pound bags.

Mixes

The same feeders that dispense sunflower seeds will handle the mixtures sold as "wild bird food." These combinations vary widely in cost, depending on the proportions of ingredients and the size of the bag. The five-pound bags of seed available in supermarkets are more expensive than comparable mixes at feed stores and garden centers.

What else besides sunflower seeds is in these mixes? Ingredients — and proportions — vary, but most contain some variety of millet, rape seed, sorghum, and canary seed. The advantages of a commercial mix are its convenience and appeal to a variety of birds.

Those who are so inclined can devise their own mixes, usually at lower cost, to appeal to particular species or in an effort to exclude "undesirable" birds. You'll probably have

Sunflower Seeds

There are two kinds of sunflower seeds. The big striped ones that we grow are used for their showy flowers and for human confections. The much smaller black seeds usually sold for bird food are the kind that are pressed for oil.

If you feed sunflower seeds, you'll have large piles of hulls under the feeders in the spring. Don't toss these into the garden or compost pile. They have a substance that inhibits the growth of plants.

to resort to a feed store to get some of the ingredients. The only common seed that I've seen at a store stocking bird supplies was yellow proso millet, the tiny round seed that forms the base of many mixes, and it was priced too high to be practical to use in your own mix. The procedure of formulating your own mix is so complicated that most of us who feed birds use a ready-made mix and feed special foods like thistle and sunflower seed separately. That eliminates finding a source for mix ingredients in bulk and also the need for storage space for them.

Cracked Corn

In some stores that sell bird supplies, you'll find cracked corn, popular with a large number of birds, especially game birds. You'll save from 50 to more than 100 percent by buying it in bulk at a feed store. You pay a tremendous amount for the convenience of getting it in the smaller amounts at a wild bird supply department.

In roughly the same price category at feed stores, you can buy scratch feed, a mixture of cracked corn and some other grain, usually wheat. It comes in three grades: fine, medium, and coarse. Choose according to the size bird you'd like to attract with it. One disadvantage to keep in mind when dispensing cracked corn or scratch feed: it spoils rather quickly in wet weather. You should dispose of feed not cleaned up in a short time, to avoid sickening the birds. Either feed very small quantities during wet weather or withhold it altogether, substituting other foods. Another disadvantage is that cracked corn is liable to attract swarms of crows.

Peanut Hearts

Peanut hearts, the embryo of the peanut, are a by-product of making peanut butter. At one time they were a fairly standard feeder item, but their cost has reduced their popularity. They're enjoyed by most birds that eat nutmeats. They have a tendency to spoil rapidly, but that's not really a major objection since they also have a tendency to disappear fast. You may want to use them for an occasional treat, if not for daily fare. Some of the more expensive mixes labeled "high energy" contain peanut hearts. Such mixes are intended for use in severely cold weather. Safflower seed is sold at approximately the same price as peanut hearts. Although it, too, is a good food, its cost makes it little used.

Eggshells

One leftover you can safely offer birds, because calcium is so important for them, is eggshells. Toasted lightly in the oven, then broken up or crushed, they can be dispensed with other foods or included in homemade suet cakes. It's believed that serving eggshells to jays will prevent them from robbing eggs from other birds' nests. During nesting time especially, it's a service to the birds to provide them with eggshells.

The healthiest and most natural food for a hummingbird is flower nectar.

Suet

"Bird" suet is widely available in northern supermarkets in the wintertime. Come warmer weather, it usually disappears, so you'll have to rely on an honest-to-goodness butcher shop, seed-suet cakes, or the suet you've providentially stashed in the freezer. Suet used to be free, like soup bones, but now you have to buy it, although it usually isn't very expensive. We always have a considerable amount on hand because we raise our own beef. Be sure to demand suet (you've already paid for it, in the hanging weight!) if you buy beef by the side. If you buy seed-suet cakes, you'll find that they range in price depending on weight. Some of them are self-contained, ready-to-hang units complete with perches.

Hummingbird Food

On the subject of filling your hummingbird feeder, experts disagree on what to use — sugar or honey — and in what proportion to water. Some say there aren't any vitamins or minerals in sugar, but their opponents argue that honey can more easily ferment and cause various problems. The recommended ratio of sugar to water is one to four. With honey it varies from three to nine parts water to one part sweetener.

Commercial nectar is a concentrated high-energy sugar compound simulating flower nectar. An accompanying vitamin supplement has been formulated to counteract deficiencies.

Whatever you use, make sure feeders are cleaned on a regular basis. And keep ants away; hummingbirds won't use a feeder that has attracted ants.

The Best Food for Hummingbirds

Planting the flowers that hummingbirds love is preferable to offering them artificial nectar. To learn how your garden can nourish these delightful visitors naturally, *see* Chapter 6.

GIVING BIRDS PEOPLE'S FOOD

Before going on to some of the non-wild-bird foods that can be fed to birds, let's consider some of the concoctions that people prepare in their kitchens especially for the feeders. In the last chapter I described the bacon grease, peanut butter, and/or cornmeal mix that's used in log feeders. Rendered suet mixed with cornmeal, peanut butter, or cracked corn can be poured into molds (tuna fish cans, margarine tubs, and the like) to harden. Feed it in suet holders or pour it before it hardens into such impromptu feeders as coconut shells. Sometimes raisins, currants, and brown sugar are added. Try your own combinations.

Leftovers

Although you can feed some leftovers to birds, don't give them bread or other foods made from white flour. These often contain additives, sugar, the wrong kinds of fats, and other ingredients that could be harmful. They can also attract unwelcome four-footed visitors.

Because kitchen scraps are likely to spoil quickly, especially in warm weather, put out only small quantities at any given time until you see what takers you have. Don't forget that some of these scraps, fed separately, may keep starlings, grackles, blackbirds, and house sparrows from monopolizing feeders. Try cooked potatoes in any form, sweet potatoes, cooked rice, pieces of fruit, leftover cooked or ready-to-eat cereal. Always be sure that scraps are in sizes appropriate to birds or else securely held in place so that they can be consumed a little at a time. Pieces that are too big will be dropped. Not only will that make your yard look trashy, it's bound to attract unwanted guests.

Among the kitchen waste items, don't forget seeds: apple seeds (you can serve them core and all), melon seeds, pumpkin and squash seeds. If the seeds are really large, putting them through a meat grinder will make them much easier for small birds to handle. If you dry them when you have excessive amounts, you can serve them later.

Speaking of waste seeds, we store hundreds of bales of hay in our barn for winter feeding of livestock. Over a few months, a great amount of chaff accumulates on the floor. It makes a welcome treat for the birds in early spring when natural food is scarce. I just sweep it out of the loft into the paddock. The ground feeders get to work on it quickly.

Grit Needed

Remember that birds don't have teeth and they need grit to help them digest food. You can offer coarse sand in the feeder occasionally to help supply it, especially during times of snow cover. Crushed oyster shell is frequently recommended, but it's hard to find, even at many feed stores. Bits of old mortar are attractive to birds, and so are coal ashes. Both probably supply needed minerals.

Selective feeding will attract the birds you prefer. Many birds like nutmeats, and some that prefer insects will eat fruits and berries. These are expensive unless you can collect wild ones and hold them in the freezer or in dry storage for occasional treats. Grain foods attract the greatest variety of birds. Color is important to birds in selecting their food. You'll recall that white bread and white suet will attract birds to brand new feeders. Many birds that like black thistle seeds are also attracted to rape and sunflower seeds, both of them black. Peanut butter and nutmeats seem to be the only brown foods birds really relish. (Brown breads, unfortunately, appeal to them much less than white ones do.) Most birds prefer red fruit and berries to yellow, and hummingbirds go most readily to red flowers.

-4-

The Water Situation

Common grackle

ODDLY ENOUGH, many people who want to attract birds neglect to provide water for them. They may have multiple feeders, which they conscientiously stock year in and year out; yet they fail to furnish water, which is as important to the birds and sometimes harder to find than food. There are, in fact, birds that will ignore all your efforts to entice them to food delicacies, but they'll readily visit a birdbath to drink and disport themselves with abandon.

To provide water is the least expensive method of attracting birds. Oh, you *can* go to a lot of trouble and expense, but it isn't really necessary. The amounts needed are small and, most of the time, serving is simplicity itself.

I confess that for a long time I was guilty of ignoring this simple, effective method of attracting birds. We live in a rural area laced with brooks and ponds. A wide trout brook runs through our land behind the back pasture. Most winters it freezes fairly early, despite the fact that it's fast and moderately wide, with deep pools here and there. The problem didn't strike me as particularly significant. There are, after all, stock tanks for the horses and cows, kept at 40°F all winter by floating heaters. It took me a while to discover that garden birds don't like to drink from stock tanks. One reason it came as a surprise is that our barnyard flocks of

Water can be the cheapest and most effective way to attract birds.

Rhode Island Reds, Buff Orpingtons, and Australorps fly up to the tanks for refreshment at least as often as they use their own facilities. I had also occasionally noticed a sparrow or other small bird drinking at the stock tanks. Every so often (quite infrequently) we found a dead bird, domestic or wild, in the water.

BIRDBATHS

Birdbaths seem to get the most business when the feeder is busiest, which isn't surprising. You want a cup of coffee or a glass of milk with your lunch, too. Grackles sometimes dunk their food before eating it. If they consider a piece of bread too hard, they'll carry it to the birdbath.

That kind of behavior can make a mess in short order, but it's not as bad as another of their habits. They often dispose of various bits of trash, including fecal sacs from their nests, in birdbaths. To their credit, they sometimes eat — or simply remove — objects they find in the water. Other dunkers include starlings, house sparrows, and red-winged blackbirds. It's usually baked goods that get the water treatment.

Among the birds likely to visit birdbaths are some of the particularly colorful ones and some that seldom visit feeders. Included are bobwhites, bulbuls (exotic immigrants to southern Florida), indigo buntings, brown thrashers, and cedar waxwings. Meadowlarks are attracted to lawn sprinklers, and white-winged doves will come to surface water.

Shallow Water

Most of the birds that come to feeding stations prefer to drink — and bathe — in very shallow water. If you've observed songbirds drinking at a brook or some other natural water source, you'll have noticed they generally use the edges. The majority of garden birds prefer the water to be no more than 2½ inches deep. Even large birds like jays and grackles are highly suspicious of water more than four or five inches deep. All of them prefer a vessel that *slopes* to the deepest point. Moreover, its surface should be rough so that they can easily keep their footing.

These modest requirements are very easy to fulfill with a run-of-the-mill clay birdbath — a shallow bowl and separate pedestal base. The bowls are also sold separately for two reasons: the tops can be used alone, on the ground; they are also breakable and somewhat more prone to damage than the bases. Decorated birdbaths of clay are slightly more expensive. Concrete birdbaths are competitive in price with clay ones.

That's the bare beginning. Anyone interested can find baths ranging from the tasteful to the bizarre, made from a wide selection of materials and priced accordingly.

Safety

The use of the pedestal-type birdbath is recommended wherever there are cats because birds with wet feathers fly poorly; besides, they frequently become so engrossed in their bathing activities that they aren't as alert as they might be. The bath itself should be in an open area — on a lawn, for example — so the birds are less likely to be caught by surprise. In placing a birdbath, make sure there are trees or sturdy bushes nearby where the birds can perch and preen after bathing.

Birds need to preen their wet feathers after bathing, before they can fly well.

OTHER WAYS TO PROVIDE WATER

Containers other than birdbaths can be used for water. The larger sizes (say, 6 to 12 inches in diameter) of earthenware saucers made for potted plants work fine. Even with only half an inch of water in them, birds will find them quite satisfactory. Robins, catbirds, orioles, and grosbeaks are among those that find such saucers adequate for bathing as well as drinking.

Garden pools can be constructed simply and attractively, beautifying the garden and satisfying the birds at the same time. Keep them shallow, and sloping. If there is dry space around the edges, so much the better. Birds are very cautious when entering water to bathe, even if they are regular visitors to the site.

Besides ordinary birdbaths, basins, and saucers, some birders report that natural stone basins that hold only a small amount of water also attract the birds. Such natural basins are easy to clean with a stiff broom. Incidentally, one warning about wading pools: Friends of ours came home from a weekend at their beach cottage to find a catbird drowned in their children's plastic wading pool. Apparently it had misjudged the water's depth and was unable to get out.

How often any birdbath must be cleaned depends on what birds use it and where it's placed. Most people prefer to position the birdbath some little distance from a feeder and any overhanging branches so that it'll stay cleaner.

Provide a Secure Footing

Plastic or metal containers can be used if you place bricks or stones on the bottom, to provide a rough surface. Pebbles or coarse sand would work just as well, but either is a nuisance when it's time to clean the container.

FOUNTAINS

Among the refinements often suggested for birdbaths are fountains. When I priced them, out of idle curiosity, at a garden store, I found that very simple but handsome ones were not inexpensive. That had a dampening effect on my enthusiasm, although in theory the idea appeals enormously to me.

In the meantime, however, I remind myself that the *sound* of water may attract more birds than would ordinarily come to the birdbath. I may be enamored of a fancy fountain, but the birds couldn't care less about its aesthetic attributes. And there's a virtually cost-free solution to the situation that's very easy to manage: a drip will serve. Water dripping into a birdbath at the rate of about a drop a second is considered irresistible to birds. There are various ways to arrange the attraction.

There's the garden hose, for one. It doesn't make an entirely satisfactory arrangement, but it does make it easy to clean and fill the bath. It looks a little peculiar — or careless — and you'll have to move it every time you mow the lawn. I've heard of people getting around this problem by complicated plumbing intricacies — burying hose or pipe and then running it up to the basin at the pedestal. Personally, I can't face the idea of extensive excavations just for a drip. For a proper fountain, okay, but not for a drip.

A Cheap Drip

There's another way to go about the drip project. It's not necessarily a perfect solution from every point of view, but try to keep an open mind. For one thing, the mechanism can be camouflaged. It *should* be. Here's how we did it. My husband, Bud, took a three-pound coffee can. Like most coffee cans, it was colorful — bright blue, with red accents. Such lurid colors are apt to alarm birds, and they don't do much to improve the premises, either. Bud sprayed it flat black. Dull green or brown would be okay, too. He fashioned a wire handle for it and punched two holes near the top of the can in which he fastened the wire. Then he punched a tiny, nearly infinitesimal, hole in the bottom of the can, slung the handle over the conveniently overhanging limb of a crab apple tree, and filled the container with water. What resulted was a steady stream, albeit small, instead of an alluring *plink, plink*.

An ingenious fellow and not one to be foiled by a mere coffee can, he took a sheet metal screw and a rubber washer and inserted the screw from the outside through the washer and the hole in the coffee can. With this simple twist of the screw, the drip can be regulated, creating the requisite *plinking* sound. There's no need to take a screwdriver to it, either. Since the screw protrudes through the bottom of the can, grasping it and turning it will suffice. It's worth noting that the drip will require attention the first couple of days because the washer swells very gradually as it becomes sodden.

As far as we know — we didn't mount a twenty-four-hour watch — it took the birds two days to discover their new recreation center. The first visitor we noticed was an adolescent robin. He enjoyed himself immensely, splashed vigorously, and then retreated to the crab apple tree to rearrange his feathers. The birds regularly congregate in the area, part of our formal garden, and merely hop or fly short distances away when we're mowing, weeding, or simply visiting the place.

The drip is likely to attract birds that might otherwise ignore your birdbath. Warblers of various kinds, depending on season and area, flycatchers, and thrushes other than robins may be among the visitors. The sound of the water gets their attention, and its movement may also enhance its appeal.

WHEN WATER IS SCARCE

Hermit thrush

What happens if you don't provide water? If the birds stick around your place despite this oversight, they're more likely to eat your small fruits and berries to get moisture. They'll "bathe" in wet foliage after a rain. Thirst is probably regulated by the birds' diet. Goldfinches, primarily seed eaters, need a lot of water, while birds such as bluebirds, which eat primarily fruit and insects, will not need as much.

In naturally dry areas, water may be *all* you need to attract birds. In other areas, however, the birdbath will get more of a workout at some times than at others. During times of

How Birds Bathe

It's entertaining to watch the different approaches various birds take to bathing. Hummingbirds, as you might expect, prefer to perform the rite "on the wing." If you have a fountain or use a lawn sprinkler, you may see them flying through it, just as they fly through the mist of waterfalls. Phoebes and kingbirds — seldom feeder visitors — like the same method.

Flycatchers fly down to the birdbath, smack the water soundly with their breasts, and then fly to a perch. They often repeat the performance. Wrens, titmice, and chickadees are in and out quickly. Cedar waxwings usually arrive in flocks, to feed or to bathe. They'll wait their turns at the birdbath, but once in it they're every bit as exuberant as catbirds. Cardinals prefer to bathe alone. Chipping, white-throated, and song sparrows can be described as compulsive bathers. The song sparrows are so fond of the activity that they try to drive other birds away. Starlings are so obstreperous in their bathing that the bath often must be replenished (and scrubbed!) after they finish their ablutions. You'll probably see house sparrows enjoying the shower that the starlings create by their splashing. Warblers, juncos, thrashers, and orioles will all be conspicuous frequenters of the birdbath.

sparse rainfall, for example, expect record attendance. During the fall months, water may attract more migrants than your feeding station does (because of the abundance of natural foods). Birds will use the water about three times as often for drinking as for bathing. In the summertime, they will bathe most often on cool days. Evidently they're not fond of warm water for the purpose and, of course, water warms up quickly in a shallow container. Birds seem to feel a need for bathing most at molting time, in August and September. Presumably it has a soothing effect on skin made sensitive by the molt.

Fresh Water

Black-capped chickadee

Keeping water fresh and abundant isn't much of a chore during warm weather, but those of us who endure cold winters have a problem in supplying the birds when the temperature drops below freezing. If there's snow cover, the birds will get along fine. Some of them — downy woodpeckers, crows, juncos, white-throated sparrows, and black-capped chickadees — will even bathe in snow. But when the ground is bare and natural water sources are frozen, the birds may be in desperate straits.

There are various solutions to the problem. One answer is to put out *hot* water during freezing weather. But the container will be a problem. Clay and concrete baths will break if we try to keep them operating in winter by such means. The shallow black pans made by Fortex are splendid for bird watering in freezing weather. They're made of virtually indestructible, heavy, rather soft neoprene-type plastic. A solid blow to the container will remove ice without damaging Fortex in the slightest, and you may pour boiling water into it with aplomb.

The highly respected John K. Terres in *Songbirds in Your Garden* (New York: Thomas Y. Crowell, 1968) speaks of the birds standing in hot water in extremely cold weather, apparently warming their feet and enjoying the rising steam. I feel diffident about questioning such an authority, but that situation would make me extremely nervous. With no evidence to support my conviction, I'm persuaded that such a footbath might actually damage the birds' feet, perhaps inviting frostbite. John V. Dennis in *Beyond the Bird Feeder* (New York: Alfred A. Knopf, 1981), for example, reports that starlings that bathed in a heated birdbath when the air temperatures was -10°F promptly froze to death. In the same book he states the startling observation that, in general, birds are very active bathers when the temperature registers from 23°–26°F. Dennis asserts that birds bathe regularly in winter to keep warm: proper care of their feathers helps to insulate them from the cold.

WATER HEATERS

If you will be away from the house and unable to provide fresh water in freezing weather when there's no snow cover, you might want to consider using an immersion heater in your watering facility. They're available both at stores that sell poultry supplies and those that sell wild bird supplies. Relatively inexpensive to operate, they maintain a fixed water temperature of 50°–55°F. They can be used in containers that don't normally make good birdbaths — say, a metal or plastic dishpan, six to eight inches deep. Put the heater in the bottom with a layer of bricks or stones covering it to provide the correct water depth — about 2½ inches — for songbirds. Friends of ours use a large pail instead and float a block of wood on the water. Most of their winter visitors perch on the rim of the pail to drink, but a few of the smaller birds use the wood block itself as a perch. We use floating thermostatically controlled heaters in our stock tanks in winter, but the tanks themselves have now been placed in the barn. If they were outdoors, I'd float lengths of plank in them to give the birds a perch and an island of safety.

If you decide to heat water for the birds electrically, make sure you use a UL-approved heater *designed* to be immersed in water. Make sure your extension cord is of the outdoor, weatherproof variety. Don't try to cut corners and concoct a Rube Goldberg heater that may be dangerous to you as well as to birds and other animals.

OTHER TYPES OF GROOMING

Birds engage in various other kinds of "bathing." While these have nothing much to do with water, they do seem to be related to feather care. Probably everyone has observed domestic chickens luxuriating in dust baths. If the procedure doesn't actually rid them of lice and mites, apparently it relieves discomfort. Chickens and their relatives, all of which are considered game birds — grouse, partridges, pheasants — are addicted to dust bathing. Indeed, they are offended by the whole idea of bathing in water. House sparrows are the only birds common to our gardens that are similarly inclined.

Canada goose

Ring-necked pheasant

You may have come across the little depressions they make in bare earth.

Birds can sometimes be seen bathing in smoke or steam. A more commonly observed form of bathing among birds, referred to as air bathing, is apparently stimulated by the sight of water and/or the sight of other birds bathing in water. While stationed on a branch, the ground, or some other dry surface, the air-bathing bird goes through all the motions of bathing in water. He then proceeds to preen his feathers, just as if he had become thoroughly soaked.

All of these bathing behaviors are necessary to keep the birds' feathers in proper condition. Animals, we may assume, are motivated to keep their coats — of whatever material they may consist — in a proper condition to protect themselves. Horses roll to relieve itches resulting from hot hides, insect bites, or shedding. They'll resort to wet places for a nice layer of dirt to assist thin summer coats in warding off fly bites. Pigs protect their virtually hairless skins from too much sun with a covering of mud. Dogs and cats groom themselves relentlessly. Just so, birds groom their feathers both to keep them in top flight condition and to relieve discomfort. They preen after their baths, whether of water, air, or sun. Most birds have an oil gland above the base of their tails, that they massage with their beaks; they then run their beaks through their feathers to waterproof them.

Anting

One of the really peculiar behaviors associated with grooming is the procedure known as "anting." This is not commonly observed, but it's something we can be on the lookout for. Most of what is known about anting has been reported by conscientious amateur birders, not by professional ornithologists. (We're useful to science as well as to the birds if we take them seriously and record our observations carefully.)

Both active and passive forms of anting have been recorded. In the passive variety, the bird plunks himself down in the midst of an anthill and permits — or invites — the insects to stroll through his plumage. In the active form of anting, the bird picks up an ant in his beak, passes it through his feathers, and then either discards it or eats it.

Anting has been observed in blue jays, starlings, catbirds, robins, house sparrows, and juncos. Starlings take the prize for entertainment value in this category. Apparently convinced that there aren't enough ants to go around, they'll quarrel over a single individual. Once actually in possession of an

A robin will "bathe" in an anthill.

ant, they are apt to become so engrossed in their activity, so eager to reach the unreachable itch, that they fall all over themselves. Starlings are little noted for grace or dignity.

It's theorized that the formic acid secreted by ants kills parasites that are irritating birds' skin. Anting, by the way, is not confined to birds. This strange behavior has been observed in two bird enemies, gray squirrels and domestic cats.

Dedicated professional ornithologists, as well as amateurs, have often waited for years for the chance to observe birds in the process of anting. Apparently, the activity hasn't been observed in Florida, New England, or the Pacific Coast region. Whether it doesn't occur in these areas or simply hasn't been observed and reported isn't known. Anting has most commonly been observed on anthills or in favored sunning sites, from mid-May through August, in periods of high humidity.

Related, probably, to anting is the practice of using other materials to groom or soothe the skin. What seems to be common to the items used is a heat-producing quality. The birds are thought to stumble on appropriate material through the process of trial and error. Among the articles recorded by observers are pieces of the rind of limes and other citrus fruits, sumac berries, chokecherries, apple peel, green English walnuts, and pieces of raw onion, to say nothing of moth balls, mustard, vinegar, pickles, and beer. Blue jays (wouldn't you know it would be blue jays!) have been photographed using still-burning cigarette butts. Now, there's a heat-producing agent with a vengeance.

-5-

Providing Nesting Boxes

JUST AS WATER ATTRACTS some birds to your yard, shelter will attract others. Providing nesting boxes is particularly important for city and suburban folk who want birds around. Difficulty in finding natural shelter near the food and water sources you supply may tempt birds to look elsewhere for a more promising environment. If you can provide a place for birds to nest, you'll have the pleasure of seeing them frequently at close range and the advantage of allies in the war against insects. For more specifics on nesting boxes, see the chart on page 57.

A tufted titmouse will accept a nesting box.

About fifty species accept birdhouses; of that number, thirty-five regularly do so. For the most part, the birds that use nesting boxes are mighty eaters of insects. Under natural conditions they'd be nesting in holes in trees or fence posts or telephone poles. You can provide artificial holes for them. With just a little attention to detail, you'll be able to attract specific birds and exclude those you find less desirable.

The birds that accept birdhouses most readily are cavity-nesting birds. Especially in city and suburban yards, natural cavities are hard to come by. Any trees old enough to have developed cavities of the right size are usually treasured by their owners. Tree surgeons will very likely have filled any cavities. Not much likelihood of ancient fence posts, either, and the utility companies replace poles regularly. That doesn't leave much for instant bird accommodations.

Red-bellied woodpecker

CHOOSING THE SITE

So if you want nesting birds, it's up to you. There's a little more to it than just hanging up a birdhouse. The site should be chosen carefully. Birdhouses hung in or fastened to trees should be clear of the main trunk and where sunshine can reach them (although houses for forest species, such as woodpeckers and chickadees, must be mounted directly on trees). One of the advantages of positioning new birdhouses in early autumn before the leaves fall is that you can make sure their locations won't be too dark and gloomy. All growing things need sunshine, including baby birds.

Features you can easily provide will attract birds and make your birdhouses desirable residences. Ventilation holes drilled near the top of the sides will provide fresh air and keep the interior from being totally dark — a bird might spook if it flew to the entrance hole and found the insides pitch black. Wood chips or shavings on the bottom of the box will be welcomed by the bird, but sawdust isn't generally recommended. For chickadees and small woodpeckers, completely fill the box with shavings and let the bird make its own cavity.

Certain considerations in placement encourage birds to use the boxes provided. Mounting the boxes on poles discourages predators, and a clear flight path to the entrance is desirable. However, the entrance hole shouldn't be exposed to windy storms, which are mostly from the east. Facing the hole to the south, southwest, or west is best. The box should be vertical or slanted slightly toward the ground to keep it from flooding.

Rodents

Rodents are as destructive to nesting birds as cats. Squirrels' acrobatic abilities are awesome, as anyone who has tried to keep them out of feeders will testify. Rats are another problem. They'll eat not only bird eggs and nestlings but sometimes, at night, the parent bird incubating the eggs. Rat control is one of many reasons to protect owls. The barn owl, which will dine on mice as well as young rats, is often considered the most useful of all our owls in this respect.

Don't create a bird slum — four to five houses per acre is about right. Even the purple martin would prefer a lower density than is commonly offered.

TIMING

If birds ignore the houses you've installed for them, be patient. A brand new house may be viewed at first with deep suspicion. Once it's weathered a bit, birds are more likely to accept it. This is another reason why fall is an excellent time to erect new houses.

Don't forget that nesting time varies not only with the bird species in question but with geographical location as well. For same-year occupancy, houses ought to be in place no later than January in the southernmost tier of states, February in the middle tier, and March in northern states and Canada.

SAFETY

Life is more dangerous for nestlings and fledglings than for mature birds because they're less able either to recognize or to escape enemies. The young of most cavity nesters (the typical users of birdboxes) are born naked and helpless; to survive they need all the aid we can give them. If only one or two of a brood survive to maturity, that's enough to maintain a species. We can help by offering protection against predators in the nesting facilities we supply.

You will, of course, want to make bird housing secure, protecting the birds from cats, raccoons, and snakes. Here are some tips:

◆ If the house is mounted on a post that cats or squirrels can climb, wrap the post with sheet metal
◆ Place 2 inches of 4-inch stovepipe around the pole to deter raccoons, cats, and snakes
◆ Mount a cone-shaped metal guard beneath the house (purchased from stores that handle bird supplies or custom-made by a sheet metal fabricator)
◆ Grease the pole
◆ Attach a 3-inch collar of galvanized metal around a tree trunk
◆ Birdhouses hung in trees 5 feet or more above the ground are fairly safe from cats

Housekeeping

Housekeeping is necessary in birdhouses; removal of the contents of the house at the end of the nesting season will cut down on parasites such as mites and lice and will discourage mice from nesting there. Clean between broods, too, removing infertile eggs and dead nestlings, because otherwise the birds will make a new nest on top of the old one, and, being higher, it will be more vulnerable to predators. A mesh platform of ⅜-inch hardware cloth at the bottom of the box will let blowfly larvae drop through and prevent them from parasitizing the birds. Some birders recommend removing the nests of any "undesirables" (such as starlings and house sparrows) that invade houses; others suggest blocking the entrance holes until the species you've targeted arrives on the scene.

Make Your Own Predator Guard

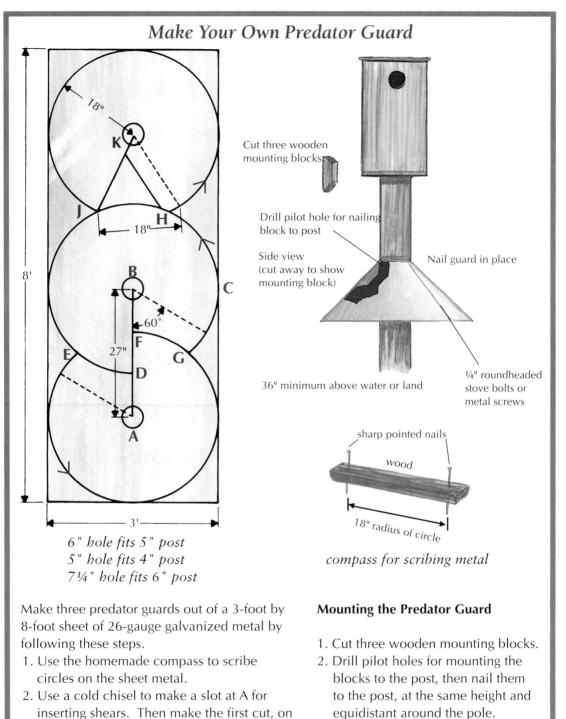

8'

18"

K C

J

18" H

8'

B

C

60°

F

27" G

E

D

A

3'

6" hole fits 5" post
5" hole fits 4" post
7¼" hole fits 6" post

Cut three wooden mounting blocks

Drill pilot hole for nailing block to post

Side view
(cut away to show mounting block)

Nail guard in place

36" minimum above water or land

¼" roundheaded stove bolts or metal screws

sharp pointed nails

wood

18" radius of circle

compass for scribing metal

Make three predator guards out of a 3-foot by 8-foot sheet of 26-gauge galvanized metal by following these steps.

1. Use the homemade compass to scribe circles on the sheet metal.
2. Use a cold chisel to make a slot at A for inserting shears. Then make the first cut, on line A-B.
3. Cut on the solid lines in this order: CD, EF, GC, small circle around B, small circle around A, HJ, JK, small circle around K, KH.
4. Note the size of holes to cut for various sizes of poles.

Mounting the Predator Guard

1. Cut three wooden mounting blocks.
2. Drill pilot holes for mounting the blocks to the post, then nail them to the post, at the same height and equidistant around the pole.
3. Form a cone around the post, above the mounting blocks, and hold the cone's shape with ¼ inch round-headed stove bolts or metal screws.
4. Nail the cone to mounting blocks.

(From *Homes for Birds*, U.S. Department of the Interior.)

Two More Homemade Predator Guards

This simple predator guard is a sheet-metal cylinder placed around the post holding the bird house. The cylinder should be at least 18 inches long and placed high enough on the pole so that cats can't spring over it.

This predator guard is simply an extra piece of 1-inch wood placed around the entrance, making it more difficult for a big bird's beak or an animal's paw to reach the nest inside. Metal circlets can be placed around the entrance holes to prevent squirrels from enlarging the entrance hole. This can be dangerous to the birds if any sharp edges remain.

BUILDING TIPS

As a considerate host you should do your best to prevent overheating, chilling, leaking, and condensation. Keep the following tips in mind.

Materials

Exterior-grade plywood or ¾-inch cedar is the best bird-house wood; it lasts a long time. Some books recommend PVC pipe, but they're in error; it heats up quickly. Metal and plastic usually aren't good because of heat buildup, so coffee cans and plastic jugs are also poor choices. But there are two exceptions to this rule. Aluminum purple martin houses are less likely to topple than heavier wooden ones, and metal wood duck houses offer protection against raccoons.

Painting

Use either stain or a brown, tan, or gray paint for the exterior finish, but make sure you don't use creosote or lead-based paints. White is recommended for martin houses only because they're large, compartmented houses sometimes made of metal and wholly out in the open, and white helps to keep the temperature down. The interior of birdhouses should be left unfinished.

Drainage and Ventilation

For proper drainage, the roof should extend at least 3 inches beyond the front of the house. Flat-roofed houses should have a ⅛-inch-deep drip line parallel to the face. And put ⅜-inch holes in each corner of the floor.

The box shouldn't be too flimsy — the heat in a thin-walled box can kill young birds. Wrens that nest in milk-carton or tailpipe birdhouses and other similarly weird places may have such a problem.

PURPLE MARTIN HOUSES

This four-story purple martin "condominium complex" should be erected in an open field.

The most conspicuous of all birdhouses are those designed for purple martins. The long tradition of providing homes for martins in this country started with the Indians, who hung gourds around their villages for the birds. You could do that, of course, but there's something about a proper martin house that appeals to many people.

Nineteenth-century England pioneered the regular use of martin houses. Homeowners like them because of their impressive appearance and the supposed value of martins in keeping down mosquito populations. It might be well to be aware that martins and mosquitoes are not necessarily active at the same time, so the effect of one upon the other is probably minimal, despite the claims of martin-house manufacturers. A bat house (there *are* such things) might be more appropriate.

Before you rush to your workshop, ship off an order, or visit the nearest bird supply house, it first would be prudent to discover the probability of martins actually moving in. If you live in a heavily wooded area, forget it. Like other swallows, they like plenty of open spaces, and there should be water nearby. Your best bet is to inquire around the neighborhood to see if any martins occupy the large edifices you're bound to see.

Make Your Own Martin House

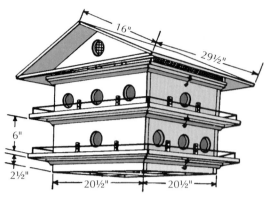

This 16-apartment martin house is made up of four separate parts that are linked together. The parts are the roof, the two stories, and the foundation.

Use ¾-inch lumber for the walls and floors, with the roof and interior partitions made of ½-inch wood. The porches are 3 inches wide, with a ½-inch dowel serving as a railing to prevent the young martins from falling to the ground. The central cross of the frame is made of double thicknesses of ¾-inch oak, and the rest of the foundation is made of ¾-inch pine. The oak central cross is attached to the pole with four heavy angle irons.

This may sound overbuilt, but it is needed when the house, weighing some 65 pounds, is placed atop a pole 10–20 feet in height. Details of the house include a central air shaft and an airy "attic," to aid in cooling the house, and a cove molding around the underside of the roof and each story to help hold the parts together. Hooks and screw eyes fasten the units to each other.

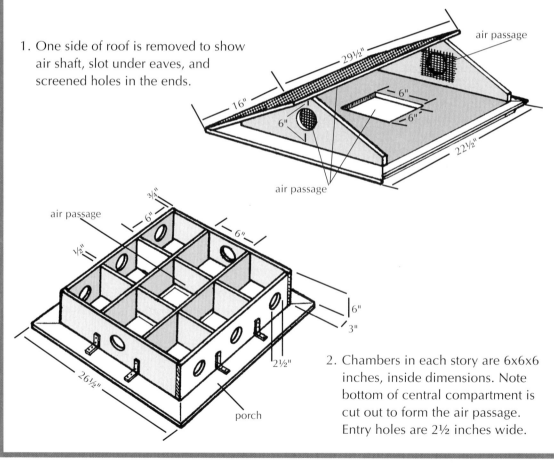

1. One side of roof is removed to show air shaft, slot under eaves, and screened holes in the ends.

air passage

air passage

air passage

porch

2. Chambers in each story are 6x6x6 inches, inside dimensions. Note bottom of central compartment is cut out to form the air passage. Entry holes are 2½ inches wide.

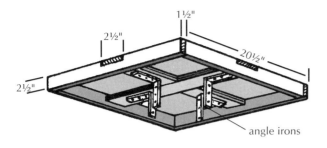

3. Sturdy foundation consists of an outer frame, an oaken cross, and four heavy angle irons.

porch

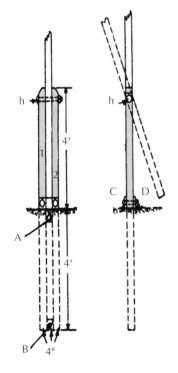

4. Porch can be attached to the house with angle irons. Note how molding (M) fits about the top of the lower story; the screw eyes and hooks (S) fasten the units together. The groove (G) is made to prevent water from draining inward.

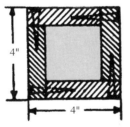

5. Here's a way to hold that martin house in the air. The supports (1 and 2) are 8 feet long and 4 inches square. They are held in alignment by 4-inch blocks (A and B) and are buried 4 feet into the ground. A long, heavy bolt (h) serves as a hinge, and the base of the pole is held in position by two hardwood blocks or iron plates (C and D) bolted together. The cross-section of the pole is shown at right, with ⅞-inch hardwood used to build it. It should be 10–20 feet long.

If you find that martins like your area, you're all set. Start small, preferably with a house that can be expanded if the need arises. Martin houses are set on posts in an open space. Because all birdhouses should be cleaned regularly, you might find the use of a telescoping pole convenient. You can get a pole that adjusts from five to fourteen feet. Its slippery surface discourages climbing predators, and it comes complete with wing nuts and clamp to fit the brackets on the bottom of ready-built martin houses. The houses themselves usually have a minimum of eight apartments and come in many different styles carefully calculated to suit the flavor of nearly any human dwelling.

All real estate is expensive, and martin dwellings are no exception. But the high cost of a martin house isn't really out of line because of its complexity. Those skilled in carpentry might like to have a go at building one.

Eastern bluebird

BLUEBIRDS

Martin houses are the most conspicuous of all birdhouses, but the most common, surely, are bluebird houses. The reasons for their popularity are numerous. Bluebirds, whether eastern, western, or mountain, are very attractive creatures, inordinately pleasing to most of us, with a song evocative of spring. Add to this the fact that they have problems, and you're bound to see champions springing up. Human population growth, especially in the eastern United States, has destroyed a large amount of bluebird habitat. Plus, the arrival of immigrant house sparrows and starlings heavily increased the competition for nesting sites, since all three birds have similar requirements. The spread westward of starlings and house sparrows has had its effect on mountain and western bluebirds, but the eastern has been hardest hit so far. The final blow came in the terrible winter of 1957–1958. Migratory birds froze to death by the thousands in the south. Bluebirds, whose diet is mainly insects and berries, were especially vulnerable. It's estimated that a third to a half of the entire bluebird population died.

Things got worse. Six dreadful winters followed. In 1963 we had the lowest eastern bluebird population ever recorded. Bird lovers, especially the North American Bluebird Society, started an active campaign to rebuild the ravaged bluebird population. Major efforts have been concentrated on providing nesting boxes specifically for bluebirds, boxes so designed that they would discourage other birds and encourage people to put them up in suitable places. This ongoing campaign has been popular and successful.

The Ideal Bluebird House

Spacing (see box) is but one of the considerations in attracting bluebirds to a house you provide. Let's examine the house itself. The North American Bluebird Society has listed the dimensions you should adhere to in order to attract bluebirds and discourage others. They're shown in the illustration. I've seen ready-made "bluebird" houses that don't meet those requirements, so take care when you go shopping. The North American Bluebird Society will send you plans or a house for a modest fee, depending on materials. It's one of the best buys available in ready-made birdhouses.

You may well ask why all the fuss about these dimensions? When they nest in tree cavities and holes in old wooden fence posts, surely bluebirds can't command accommodations so meticulously exact. But we're talking about a bird whose natural environment has been modified, whose competition has increased — and one whose numbers we're trying to increase. The dimensions are designed to suit the bluebird specifically while making the house undesirable for other species. The floor, for example, is 4" by 4" or 5" by 5". That size is adequate for bluebirds, but house sparrows find it cramped quarters because their broods are usually larger. The

Bluebird Trails

For the past thirty years, teams of volunteers all over the eastern United States have been building "bluebird trails." Selecting appropriate surroundings and requesting the owner's permission, they place bluebird houses at least 100 yards apart (some authorities recommend 400 feet). That distance is important because bluebirds have strong territorial drives and will engage in spirited disputes with one another if houses are too close together.

Bluebird Requirements

To attract bluebirds to a nesting box, remember the following:
- ◆ Erect box in late winter or early spring
- ◆ Entrance hole must be 1½ inches in diameter
- ◆ Do not attach a perch
- ◆ Floor must be 4" by 4" or 5" by 5"
- ◆ Place the box 3 to 5 feet from ground level
- ◆ Place boxes 300 to 400 feet apart
- ◆ Position box so entrance hole faces south
- ◆ Set box in the open, with a fence, tree, or shrub 25 to 100 feet away from entrance hole, for safe first flights
- ◆ Houses out in the open should be light colored to prevent overheating
- ◆ Clean out box between broods

1½-inch entrance hole is too small for starlings to use. Eliminate the perching peg beneath the entrance hole to prevent starlings and other bird marauders from landing to harass bluebirds at home. Place a ¾-inch block over the entrance hole as well so starlings can't reach in and peck eggs or young. Positioning the house only 3 to 5 feet from ground level also discourages house sparrows, which prefer higher nesting sites. And if you place the nestbox far from house and barn and plug the hole until the bluebirds arrive, the house sparrows (always around) will find other quarters. The same technique is useful for martin houses.

There are other birds considered more desirable than house sparrows and starlings that may also have designs on the bluebird houses. In all likelihood, you'll want to provide other houses for them while keeping the bluebird houses available for bluebirds. To prevent wrens from claiming the bluebird box, avoid shrubby areas, which they like. Put the box out in the open.

It's helpful if there's a fence, tree, or shrub somewhere between 25 and 100 feet away from the bluebird box, in direct view of the entrance hole. The fledglings will have a safe place to land after that first flight. Have the houses in place early in the spring. Bluebirds are among the first to arrive in nesting areas, and they get to work at once, often producing two to three broods.

The houses should be cleaned out between broods. If you see evidence of lice or mites, dust the box with 1 percent rotenone powder.

Bluebirds — or Swallows

Follow all the recommendations carefully and you *may* get bluebirds. Then again, you may get tree swallows. (We do.) But they're nice, too. You can use pairs of boxes if you have tree swallows or violet-green swallows as well as bluebirds. These species will coexist in the same territory, although two pairs of the same species will not. If you're a city dweller, however, you might as well forget about attracting bluebirds. They're far more likely to be found on farms, around old orchards, or in rural villages.

Tree swallow

WREN HOUSING

Wrens are more easily attracted than purple martins or bluebirds; in fact, they're more likely to take advantage of quarters you provide for them than any other bird. Usually,

Make Your Own Bluebird House

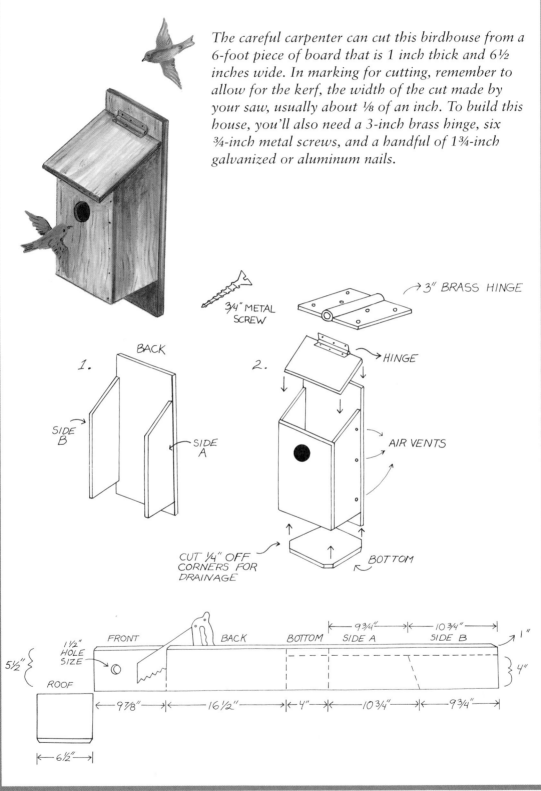

The careful carpenter can cut this birdhouse from a 6-foot piece of board that is 1 inch thick and 6½ inches wide. In marking for cutting, remember to allow for the kerf, the width of the cut made by your saw, usually about ⅛ of an inch. To build this house, you'll also need a 3-inch brass hinge, six ¾-inch metal screws, and a handful of 1¾-inch galvanized or aluminum nails.

¾" METAL SCREW

3" BRASS HINGE

1.

BACK

SIDE B

SIDE A

2.

HINGE

AIR VENTS

CUT ¼" OFF CORNERS FOR DRAINAGE

BOTTOM

ROOF

1½" HOLE SIZE

FRONT

BACK

BOTTOM

9¾" SIDE A

10¾" SIDE B

1"

5½"

4"

9⅞"

16½"

4"

10¾"

9¾"

6½"

The entrance hole to a wren house should face south.

because of competition and territorial prerogatives, it's impossible to attract more than one nesting pair of any given species to a normal-size lot. Not so with wrens. Some males will mate with more than one female, and they don't seem to trouble themselves with separating the families to avoid scandal. With their cheerful song, amusing antics, and appetite for insects, wrens are welcome tenants.

Of all the cavity-nesting birds that will use birdhouses, wrens are most likely to seek peculiar nesting locations. They've been known to use discarded shoes, tin cans, and the leg of a child's blue jeans hung up to dry.

A North Carolina wren couple built a nest in the middle of my brother and sister-in-law's brand-new double pink vining geranium. Hardly had Bill placed the plant in a hanging basket attached to the overhang just above their front steps when wrens set up housekeeping. Mama laid her six pinkish brown spotted eggs smack in the middle of the geranium. Five of the eggs hatched. While the babies were still nestlings, Dee watered the plant very carefully, almost by droplets, in hopes of preserving it without damaging the birds. Four of the five nestlings survived to fledgling size and departed. When they left the nest, Dee investigated the geranium and found that the deep cup-like structure had displaced most of the soil in the middle of the plant. She replaced it, and the plant — though scarcely robust — survived.

Wrens Can Be Particular

Even though wrens are far from fussy, every once in a while someone reports difficulty in attracting them. Doris and Frank, in Illinois, hung a wren house. They heard wrens in their neighborhood, but for a couple of years their house remained vacant. One day Doris's sister remarked that they might have better luck if they turned the house so that the entrance hole faced south. They did that, and within *half an hour* a male alighted on a nearby branch and started singing, announcing his discovery to the world. He promptly brought nesting materials, which his mate just as promptly dumped out. It would be pleasant to relate that they raised a family there, but they didn't. Apparently they simply couldn't agree on the furnishings. It wasn't until two years later that a couple in agreement on such matters set up housekeeping and raised a brood. Don't give up too soon if tenants are slow to settle in.

A Gourd Experience

My Aunt Vi in Highland Park, Illinois, tells me that Uncle Sumner liked to use gourds for wren houses. He dispensed with perches but drilled "almost quarter-sized" holes, one to the front and one to the rear (not just one, the usual procedure). Hardly was the first one hung before a pair of wrens took up residence. Both birds worked together on hauling in nesting materials at first, but after a while the male took over the heavy work and the female remained inside, apparently arranging it to her satisfaction. According to Vi, it was quite a sight to watch him trying to maneuver a five-inch twig (that's easily as long as a wren!) into the entrance hole. Once he got it inside, his mate promptly chucked it out the other hole. They used both holes indiscriminately.

Bewick's wren

Vi recommends hanging gourd homes from an old wire coat hanger snipped once just below the hook, on one shoulder line. This leaves one long piece of wire with the hook at one end. Straighten it out. Bend the snipped end to fasten to the gourd. Then you can hang it in a handy tree without needing to drag out the ladder.

Nesting Shelves

Some birds will use nesting shelves or platforms that we provide in places safe from predators and reasonably sheltered from weather. Not cavity nesters, these birds include phoebes, robins, and barn swallows. Our house, barns, and sheds are themselves part of the birds' environment, and some birds, without any encouragement from us, will find a place they consider suitable as nesting sites on a building. (Chimney swifts have become so accustomed to using chimneys as a nest site that they are no longer known to use anything else.)

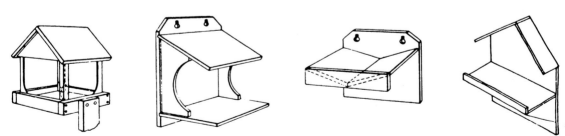

Phoebes, robins, and barn swallows prefer these shelves for nesting, rather than the closer confines of a bird house. The two shelves on the right are for hanging in areas protected from the elements.

The open barn is inviting to barn swallows.

Occasionally we have to take steps to prevent their using specific locations such as a light fixture or a door frame. But quite apart from nesting, birds depend on our buildings for shelter. In severe winter weather, areas around chimneys are particularly popular. Roofs, of course, are also used as perching and look-out sites.

Barn Swallows, Robins, and Phoebes

Barn swallows, robins, and phoebes like shelves or platforms under the eaves. Make the shelf 6 inches square and 8 to 10 feet above the ground for the robins and barn swallows; for the phoebes, it should be 6 by 8 inches and 6 to 15 feet up.

Robins, too, frequently build their nests on houses or in garages. Friends in Amherst have a robin family in their carport every year. It's no particular problem in a carport, but a garage can be another matter, especially if you want to close the door. A couple of years ago, a robin in Illinois quite appropriately chose Mother's Day to begin her nest – in a hanging basket of flowers on a deck. She raised a family of three with no apparent difficulty resulting for birds or plants. I've known people who were unable to use a door of their house for weeks at a time because robins nested above it. If you don't care to suffer that inconvenience, remove the nest at once and hope the birds will find a better place.

Other Small Birds

Pairs of tree swallows, like wrens, will nest closer together than most other birds. It's best, though, on a small lot to supply them only one or at most two houses, 100 feet apart. If you're building or buying something specifically for these two species, as opposed to letting them do what they can with bluebird houses and geraniums, a house 8 inches high with a 5-inch by 5-inch or 4-inch by 6-inch floor will accommodate either. The 1½-inch entrance hole should be 6 inches above the floor. Don't forget that all birdhouses need drainage holes and adequate ventilation.

Studies have indicated that wrens are partial to red and green houses and will use white ones only as a last resort. It's difficult to predict what birds will like in houses, but it's generally supposed that most prefer natural-looking ones. Chickadees, titmice, downy woodpeckers, and nuthatches apparently like a rustic-looking home made of wood with the bark left on. Perhaps that's because they're woodland birds. They'll also accept weathered lumber, but no bright colors, please. The same dimensions acceptable to wrens and tree swallows will do nicely for chickadees, titmice, nuthatches, downy woodpeckers, and prothonotary warblers.

Nesting Boxes for Larger Birds

For larger birds, a nesting box 14 inches high with a 6-inch by 6-inch floor and a 2-inch entrance hole 9 or 10 inches above the floor will attract great crested flycatchers, starlings, red-bellied woodpeckers, and hairy woodpeckers. On up the scale, a 16-inch-high nesting box with an 8-inch by 8-inch floor and a 3-inch entrance hole 14 inches above the floor will attract flickers, kestrels, and screech owls.

Interestingly enough, both screech owls and kestrels especially like to occupy such nesting boxes in the suburbs or the city. Though their usual prey are insects and field mice, they *do* occasionally feed on small songbirds, so they may not be as welcome in urban areas as in rural ones. The diet of both birds is the same, but they're not competitors since kestrels work days and owls work nights.

Rural dwellers who want to attract these birds should select a site 10 to 30 feet above the ground. Orchards are especially good for the owls. Kestrels like trees along the edge of a woodland or an open field.

See the chart on page 57 for more information.

Knee Deep in Phoebes

Most of us have had experiences with birds using what we consider inappropriate nesting sites. When we were building a house in the midst of a heavily wooded area, a phoebe insisted on building a nest in the floor joists. The workmen removed it daily. Then came a spell when heavy rains held up work for over a week. When work recommenced, there was the phoebe nest, with eggs in it. The activity and noise proved too much for the phoebe; she abandoned the nest.

The next year phoebes used the top of the shutter outside our son Tom's bedroom window. He was able to watch the daily progress of the brood. Another nested in the horse shelter; a third, in the goat shed. We've never built a shelf for phoebes, but we're always up to our knees in them anyway.

Simple Is Best

Most public libraries have plenty of books that give straightforward, acceptable directions for building serviceable birdhouses. Beware those books that have complicated plans and cutesy designs; it may be assumed that such accommodations are designed with humans, not birds, in mind. I've seen books that specialize in brightly colored, heavily decorated birdhouses, intended apparently to appeal to the pride of the woodworker. One book I studied actually had a birdhouse designed to look like a miniature outhouse.

Screech owl

Specific Requirements for Nesting Boxes

Ducks

Wood duck houses must be placed near water or in a swamp and raised up on a pole at least 3 feet high. The houses need ventilation holes in the bottom and bedding 4 inches deep, either wood shavings or sawdust.

Owls

Great horned owls, long-eared owls, and great gray owls will nest in chicken-wire baskets 1 or 2 inches deep supported on boards or poles. The baskets need sticks inside and should be lined with coniferous branches and a layer of twigs, leaves, and moss. The area around the nest should be fairly open for a distance of 15 to 25 feet. Screech owl boxes can be strapped to the underside of a branch; if there's no angled branch, mount the box itself at an angle.

Barred owls will nest in square or round boxes 20 to 24 inches high and 12 inches square or in diameter. Drill a 6-inch hole for an entrance, put several inches of wood chips, dried leaves, or moss in the bottom, and hang it 17 feet high in a large tree.

Doves

Mourning doves will use a cone-shaped nesting platform of ¼-inch or ⅜-inch hardware cloth placed in forked branches 6 to 16 feet from the ground.

NESTING MATERIALS FOR BIRD BOXES

We can further help the birds in their nesting activities by providing them with nesting materials. Swallows, both barn and cliff, need mud. So do phoebes, robins, and wood thrushes. You can help them by placing a shallow pan in some accessible but out-of-the-way place and filling it with soil, preferably with some clay in it to make it good and sticky. Keep it wet. The birds will be appreciative, especially if a dry spell occurs at their nesting time.

A little pile of dry twigs will be welcomed by the wrens. Feathers are popular for lining nests, especially with swallows. I've read that they prefer white feathers, but around here they make do with red and black and buff. If you make string, twine, or horsehair available, be sure the materials are no longer than six or eight inches because longer pieces can form loops and strangle the birds you're trying to assist. *See* the chart starting on page 290 for more information.

Nesting Boxes

Species	Floor of Cavity (Inches)	Depth of Cavity (Inches)	Entrance above Floor (Inches)	Diameter of Entrance (Inches)	Height above Ground or Water (W) (Feet)	Preferred Habitat Codes[3]
House Wren	4×4	6–8	4–6	1–1¼	4–10	2,7
Chickadee	4×4	9	7	1⅛	4–15	2
Bewick's Wren	4×4	6–8	4–6	1¼	5–10	2,7
Titmouse	4×4	9	7	1¼	5–15	2
Downy Woodpecker	4×4	9	7	1¼	5–15	2
Prothonotary Warbler	4×4	6	4	1⅜	4–12, 3W	3,5
Nuthatch[1]	4×4	9	7	1⅜	5–15	2
Carolina Wren	4×4	6–8	4–6	1½*	5–10	2,7
Bluebird	4×4	8–12	6–10	1½*	3–6	1
Tree Swallow	5×5	6–8	4–6	1½*	4–15	1
Violet-green Swallow	5×5	6–8	4–6	1½*	4–15	1
Ash-throated Flycatcher	6×6	8–10	6–8	1½*	8–20	1,6
Hairy Woodpecker	6×6	12–15	9–12	1⅝	12–20	2
Great Crested Flycatcher	6×6	8–10	6–8	1¾	8–20	1,2
Golden-fronted Woodpecker	6×6	12	9	2	10–20	2
Red-headed Woodpecker	6×6	12	9	2	10–20	2
Purple Martin	6×6	6	1	2½	10–20	1
Saw-whet Owl	6×6	10–12	8–10	2½	12–20	2
Flicker	7×7	16–18	14–16	2½	6–30	1,2
Screech Owl	8×8	12–15	9–12	3	10–30	2
American Kestrel	8×8	12–15	9–12	3	10–30	1,4
Barn Owl	10×18	15–18	0–4	6	12–18	4
Wood Duck	12×12	22	17	4	10–20, 6W	3,5
Phoebe	6×6	6	(²)	(²)	8–12	7,8
Barn Swallow	6×6	6	(²)	(²)	8–12	7,8
Robin	6×8	8	(²)	(²)	6–15	7

*Precise measurement required; if diameter over 1½ inches, starlings may usurp cavity.

[1]Brown-headed and pygmy nuthatches (1⅛), red-breasted nuthatch (1¼), and white-breasted nuthatch (1⅜) will all use the same box. However, the smaller opening sizes where appropriate may discourage use by house sparrows.

[2]One or more sides open.

[3]Preferred habitat codes. The numbers in the last column refer to the habitat types listed here:
1. Open areas in the sun (not shaded permanently by trees), pastures, fields, or golf courses.
2. Woodland clearings or the edge of woods.
3. Above water, or if on land, the entrance should face water.
4. On trunks of large tree, or high in little-frequented parts of barns, silos, water towers, or church steeples.
5. Moist forest bottomlands, flooded river valleys, swamps.
6. Semi-arid country, deserts, dry open woods and wood edge.
7. Backyards, near buildings.
8. Near water; under bridges, barns.

Source: Reprinted from Homes for Birds, Conservation Bulletin 14, U.S. Department of the Interior.

-6-

Gardening's for the Birds

Birds will feast on your sunflowers as soon as the seeds ripen.

YOU CAN HAVE a garden full of birds year-round without placing so much as a single feeder. The trick is in planting trees, shrubs, vines, and herbaceous plants that will provide food for the birds. So why bother with feeders at all under those conditions? For the pleasure of seeing many birds of different species close up. You'll recall that it takes a lot of acreage to support any significant number of birds. To get variety and numbers requires careful planning and careful planting. The optimal arrangement is a blend of natural foods supplied by the garden and supplemented by feeding stations. By attending to garden, water, and feeders, you can appeal to a tremendous array of birds.

As far as the birds are concerned, the very best gardens contain a combination of trees, shrubs, and lawn, as well as flower and vegetable plots. If you can add a brush pile or thicket and manage not to keep the surroundings absolutely immaculate, you'll have hordes of contented birds. (In theory, I approve of immaculate gardens. Since I seem incapable of ever quite achieving or maintaining that condition, it's comforting to realize that untidiness has its own rewards.) Songbirds live an average of three to four years (occasionally to ten) and are perfectly willing to stay in an area that provides favorable food and nesting sites, or return to it annually if they're migrating species.

The coming of colonists to this country had a great effect on bird populations. Although settlers engaged in questionable practices such as introducing new species and hunting indiscriminately for food and plumage, they opened the continent to large-scale agriculture, which provided more food for birds. Birds had had ample cover up to that time, but the largely uncultivated countryside hadn't produced enough food to support large bird populations.

Another change came with the growth of urban areas; still another, with the widespread use of agricultural machinery. Cover disappeared not only with the cutting of forests but also with the disappearance of hedgerows between fields. When farmers and landowners got going with the wholesale use of herbicides and pesticides, the birds were in bad trouble.

It doesn't have to be that way, of course. We needn't stand by and do nothing, assuming we're helpless against the forces affecting the environment. Not only *can* we do something, we *ought* to. Even from an entirely selfish point of view, we must realize that if the environment poses a threat to birds, it can't be good for us either. You can figure that if your garden — or your city or town — has a varied and healthy wild bird population, it's a pretty fair place for human beings to live. It's to the advantage of all of us, whether or not we can attract birds or maintain feeding stations, to join in efforts to protect natural habitat.

Herbs for Nesting Materials

Some of the plants in your garden may supply nesting material for birds. I don't recommend thistles or milkweed for the down that American goldfinches use in their nests, but some herbs — yarrow, rue, thyme, and pyrethrum especially — appeal to birds. Interestingly, these herbs have insecticidal properties.

Mallards may return to the same pond year after year.

Garden Flowers

Many herbaceous garden flowers supply nectar to hummingbirds: bee balm, petunias, day lilies, dahlias, columbines, scarlet lobelia, fuchsias, impatiens, red-flowering penstemons, and geraniums. Goldfinches delight in the seeds of cornflowers, cosmos, and coreopsis. Cardinals like four o'clock seeds.

Some authorities on wild bird feeding tell us about the foods and techniques we can use to "wean" wild birds away from their natural foods and get them to visit our feeders for sustenance instead. I submit that such an approach is foolish, if not downright immoral. How much better it is to provide them with a choice, right in our own yards. Each of us can maintain a minisanctuary for wild birds. Growing plants that they prefer will give us the opportunity to enjoy more species at close range, without corrupting them. They'll repay the attention by the pleasure they give us and also by consuming insects, weed seeds, and even rodents.

We started gardening in earnest, on a half-acre plot, and got Peterson's *Field Guide to the Birds* in the same year. The bird population seemed to grow in direct proportion to our advances in gardening. We counted on them as allies from the very beginning because we didn't like the idea of using pesticides. Admittedly, we failed to take into consideration the fact that the presence of birds doesn't guarantee an absence of bugs. Birds help to maintain a natural balance, but they don't discriminate between what we consider "useful" and "harmful" insects.

It seemed like a natural progression from seeing birds in the yard to adding plants that would encourage them to stick around. But that's a rather different thing than saying we planned the yard for the birds. It was, from the beginning, designed primarily for people and their pets and domestic livestock. The birds just happened. To assert their usefulness is actually beside the point. It pleased us to have them around, and most of us are delighted to find practical reasons for what pleases us.

The one time we started from scratch was nearly thirty years ago when we built a house in a clearing in the woods. We had trees, ferns, mountain laurel, blueberries, ledge, and a wet-weather brook. At first we cleared only enough trees to put in the house and driveway. Naturally enough, we saw only woodland birds — chickadees, nuthatches, woodpeckers. Gradually, over a period of several years, we opened up the area to make room for gardens and paddocks. As we got more sunshine and more open spaces, varied shrubs and herbaceous borders, the bird population became more varied. I was thrilled to see our first robin in that location.

If we'd been actively hunting a place for a little bird sanctuary, we couldn't have done much better than our present home. It's a very small farm, with about six acres of lawns, gardens, and pasture, and the rest — another three or four acres — wooded. A splendid trout brook runs through the wooded area. What at one time had been a pond is now a

A ring-necked pheasant may forage under an apple tree.

marsh. To the west of us lies a large cornfield, to the north and east are the village and farms, to the south woodlands. Canada geese visit the cornfield in the fall after the harvest. Ducks visit the brook, and kingfishers and a few mallards make their summer home there.

Pheasants parade through the side yard, under the ancient apple tree. Grouse nest in the woods above the brook, red-winged blackbirds, in the marsh. Mourning doves, phoebes, and kingbirds have their residences in the back pasture, swallows in the barnyard and front pasture. Occasional visitors include hawks and herons. Nearer the house, robins, grackles, starlings, orioles, and other common feeder visitors make their homes. Because the character of the land is so varied, it attracts birds of open land, marsh, and woodland. In the years we've lived here, we've tried to make it more appealing to us and to the birds, but by no means can we take full credit. Both resident and transient species are determined at least partly by factors over which we have little control.

PLANNING PLANTINGS

An orderly process is best in planting to attract birds. In any landscaping effort, it's sensible to start on paper rather than with a shovel, whether you're modifying an existing garden or beginning a brand-new one. The same rules apply, regardless of where you start. Each starting place has its own built-

If you add a "dusting" area
— a 3-foot by 3-foot open
space with "dust" made of
equal parts of sand, loam,
and sifted ash to a depth of
6 inches — your real
estate will appreciate
considerably from the
point of view of game
birds such as pheasants
and grouse, which like to
bathe in dust just as much
as chickens do. Of the
common garden birds,
however, only house
sparrows like dust baths.

in advantages and disadvantages. It's nice not to have to rectify past mistakes, your own or someone else's. On the other hand, it's pleasant to have the lawn in and some landscaping done. If there's a mistake, so what? You can live with it more easily than you can live with a void.

In planning plantings, take into consideration the house and any other fixed structures, such as driveways, terraces, toolsheds, and the like. Some plants that are most attractive to the birds are inappropriate near these features because they're messy. In all fairness, of course, the same objection can be made to certain plants that birds virtually ignore, such as the rose of Sharon.

Any property can serve as a sanctuary for birds. "Sanctuary" implies that the property supplies natural cover, adequate food (including supplemental food when it's necessary), and protection from natural enemies. The numbers and species of wild birds you attract will depend on the actual size of the property as well as the variety of its natural features. If you have upland green forage or aquatic plant forage, expect game birds in addition to songbirds. Flying insects will attract phoebes and, if there are wide-open spaces, kingbirds and swallows. Terrestrial insects, including Japanese beetles and their larvae, are eaten by robins, cardinals, flickers, and starlings. Aquatic animals (if you have a brook) may well bring kingfishers; rodents and various other small mammals attract hawks and owls. Depending on our circumstances, the presence of songbirds themselves may attract some birds of prey.

WHAT TO PLANT

Going Natural

Brush piles, patches of
weeds, tall grass, and dead
limbs all supply cover to
birds and to the things
they feed on. Whether or
not you're willing to
supply such amenities will
be determined at least
partly by the amount of
land at your disposal.
What's no trouble on a
large lot may be inappropriate in a small area.

It's suggested, on general principles, that 8 to 12 percent of your permanent plantings should afford food or protective shelter for birds. Let's begin with the largest and most important of these permanent plantings — trees. Some of them will provide food and nesting sites as well as shelter. And birds such as woodpeckers and creepers will be attracted to insects on your trees.

Evergreens

In colder climates especially, a variety of evergreens is useful for bird shelter and also to add color and interest to your yard after the deciduous trees and shrubs are bare. Some gardeners choose evergreen shrubbery, but if you have the space, a selection of evergreen trees adds a welcome contrast in most climates. Pines of whatever variety, find favor with most people and are equally popular with practically all birds. Their long soft needles provide shelter from severe weather, and their cones provide food. Crossbills, possessed of uniquely shaped mandibles, are peculiarly well suited to feed among the conifers; but goldfinches, towhees, tanagers, and chickadees are able to eat pine seeds, too. Pines are favorite nesting sites for many garden birds, including grackles and jays.

Besides those trees identified by the word "pine" in their common names, the family includes several other familiar conifers. Among these is red cedar, a great favorite with a large number of songbirds considered desirable by most fanciers. Red cedar is a kind of juniper and, like juniper, produces berrylike cones. Spruces, firs, and hemlocks also belong to the large pine family. Balsam firs attract large numbers of birds — finches, nuthatches, game birds, grosbeaks, chickadees, and crossbills. Finches are also quite partial to the seeds of spruce trees.

One northern conifer is a little confusing because it sheds its needles in the fall. This is the beautiful larch, or tamarack, tree. Goldfinches find its cones irresistible.

Northern flicker

Shade Trees

A large number of favorite shade trees are as welcome to birds as to the homeowner anxious to beautify his property and make it more desirable. Generally speaking, the tall, handsome trees that have long lives take quite a while to reach maturity. Even when they are young and modest in size, however, they will prove worthwhile aesthetically and a boon to the birds. The oaks, synonymous with strength, boast over fifty species in the continental United States (some authorities suggest seventy-five species as being nearer the mark). Oaks are members of the beech family and are most common in the eastern section of the country. Mourning doves, flickers, woodpeckers, jays, titmice, nuthatches, and thrashers are among the birds that feed on their acorns. The American beech, a lovely native tree in the eastern United States, produces nuts eaten by game birds, crows and grackles, jays, woodpeckers, crossbills, grosbeaks, and finches.

Perhaps best known next to oaks are the maples, whose paired winged fruits are a favorite source of food to wild birds such as evening grosbeaks, purple finches, and pine siskins. The box elder, a maple family member common to the central states, provides food for evening grosbeaks. The flowers of the horse chestnut attract hummingbirds; the seeds of the white ash attract game birds, finches, grosbeaks, and

Pruning for Birds

Pruning can create crotches in bushes and trees to provide support for the nests of sparrows, finches, and warblers. Sometimes artificial aids are better than natural conditions because they help to exclude the bird's natural enemies. Creating tree cavities by drilling 2-inch-deep holes into the heartwood, especially in rotten trees, will attract chickadees. Try to choose a place about 3 inches below a stout branch. Another technique is cutting a 3-inch-diameter branch 6 inches from the tree trunk. The branch will rot, forming a natural cavity.

cardinals. Seeds of American elms appeal to finches and cardinals. The Chinese and Siberian elms, which are resistant to Dutch elm disease, produce buds very much liked by goldfinches and Bohemian waxwings. Sweet gum, a handsome tree related to witch hazel, produces dry, round fruit full of winged seeds eaten by many songbirds.

Smaller Trees

You don't have room for such enormous trees? Let's move down in size to the birch family. These wonderfully attractive trees produce catkins, green in summer, that enlarge and turn brown in the fall. Cardinals love them. So do crossbills, finches, grouse, nuthatches, and chickadees. These birds all feed on birch buds, too. The American hornbeam, also a member of the birch family, appeals to most of the same crowd. Hornbeams are relatively small, but alders are practically shrubs, except for those growing along the Pacific coast. They're all related and produce similar seeds and buds favored by the birds that like birches.

Black gum, also called tupelo or sour gum, is another medium-size tree. Its fruit attracts bluebirds, catbirds, kingbirds, flickers, mockingbirds, robins, thrashers, and tanagers. With a track record like that, it's no wonder that the tree is popular with bird lovers in the eastern third of the country (excluding New England), where it thrives. Cottonwoods, of the willow family, are favored by ruffed grouse and evening grosbeaks.

Let's suppose you'd like some fruit trees. Peaches, apples, and pears attract birds, but plant a cherry tree — *any* kind of cherry tree — and word will travel to every bird for miles around. If you're counting on cherries for your own use, better protect some of the branches with netting. You'll get bluebirds, grosbeaks, robins, crows, thrushes, game birds, waxwings, and woodpeckers. The tree that's reputed to help protect early fruits and berries is the mulberry, a fast-growing tree whose fruit is as appealing to birds as anything you can plant. But don't plant it too close to your house or other fixed features like a patio; it *is* messy when the fruits drop. If you have the right kind of climate, you might want to grow date palms for bluebirds, mockingbirds, robins, and waxwings. Bananas will attract jays, mockingbirds, and orioles.

Ornamentals

Whether your garden is large or small, you'll surely want to find a spot to tuck in one or more smaller ornamental

Cedar waxwings will enjoy your cherries.

trees. They'll please you just as they do the birds. Try catalpa, a small tree of the trumpet creeper family, if it will grow in your climate. Evening grosbeaks think it's wonderful.

In subtropical climates, the camphor tree attracts bluebirds, cardinals, robins, waxwings, starlings, and mockingbirds. Hummingbirds come to the chinaberry. Dogwoods are popular with songbirds, and so are magnolias in their various sizes and varieties. Hawthorns, actually members of the rose family, are an extremely varied group of somewhere between 165 and 1,200 species, depending on which authority you consult. You're almost sure to find one you can grow. Their fruits look like tiny apples and are eaten by many favorite birds. Locusts appeal to hummingbirds, titmice, chickadees, and game birds. Orioles, evening grosbeaks, and bluebirds are among the birds that feed on mountain ash. Redbud attracts game birds, cuckoos, and grosbeaks. Especially in combination with dogwoods, they're breathtakingly beautiful trees.

Don't forget crab apples, which are lovely in the spring. If the waxwings discover a variety they like, they'll strip it of its fruit in a couple of days. Hopa crab apples are especially popular with them. Robins are likely to descend on Russian olives in the fall. And if you have space for a little "wild" area, let the weedy sumac move in. Fifty species of wild birds feed on it. That's reason enough to keep it around, even if you didn't get the added bonus of brilliant color in early fall.

I suppose my bias is plain. Feeders that we place for birds are very important in any program to attract them to our environs, but the garden itself provides natural food and shelter while satisfying our own aesthetic senses. That's hard to beat.

Before leaving the subject of trees, we ought to note that woodpeckers and creepers are among those birds that search out insects in *any* tree. Sapsuckers and woodpeckers make holes in trees, but many other birds (and squirrels) will use the holes to take sap or the insects trapped by the sticky substance.

Shrubs

If you're not worn out from planting all those trees, you may as well dig some holes for shrubs in the empty spaces. Or perhaps shrubs are more appropriate for your property than trees. In dry areas of the west and southwest, hummingbirds are attracted to the pink orchidlike blooms of the desert willow, the scarlet tubular flowers of the ocotillo, and the succulent white flowers of the soaptree yucca.

Nut Trees

Nut trees generally take a considerable amount of time to produce a crop, but they are valuable assets. Most birds like pecans and walnuts. Once the trees mature, there'll be enough for you, too — provided you beat the squirrels to the harvest. Chickadees, finches, sparrows, nuthatches, wrens, woodpeckers, titmice, and some warblers like butternuts and cashews. Almonds will grow anywhere that peaches will, and the trees remain modest in size. They attract cardinals, titmice, chickadees, crossbills, wrens, and warblers. Filberts, or hazelnuts, will command attention from jays, woodpeckers, and many game birds.

Bullock's oriole

Birds and Buds

You may notice that, especially in early spring when natural food is somewhat scarce, birds will eat the buds — both flower and leaf — of early bloomers like maples and fruit trees. It has been suggested that these are especially nutritious, but it's worth speculating that maybe the practice has more to do with availability than with nutritional value. It's not usually harmful to the trees. However, a friend who lives in the San Fernando Valley in California tells about a large flock of robins that discovered his apple tree in full bloom and stripped it clean. It was rare to see robins in his neighborhood. He suspects they were lost at the time of their visit.

Many fruiting shrubs appeal about equally to birds and people, a situation that may cause some consternation unless you net them for your own use or grow enough to share. We have two areas where raspberries and blackberries grow. The one adjoining the vegetable garden is largely ignored by the birds. I think the only reason they let that plot alone is that there are so many other berries growing wild behind the back pasture and around the edges of the marsh. They're undisturbed eating there, but the garden is often a site of human activity.

Our family is less than enthusiastic about gooseberries. We don't cover them and seldom see a ripe berry. The blueberries are something else again; we race the birds for the cultivated varieties. Bluebirds, robins, thrashers, and waxwings are attracted to the red currants, but we usually get enough for jelly. We grow two varieties of currants, one of which ripens much later than the other. If the first crop is adequate for our needs, we leave the second crop for the birds.

We also grow two varieties of bush cherries. One is primarily ornamental, and we leave the harvest for the birds. The other makes an excellent jelly. Both are attractive enough to earn a place in a garden strictly on their appearance; the produce is an added attraction that we welcome but don't always use. When we neglect to pick the cherries, the birds promptly take over the chore.

Various birds like elderberries. Cultivated varieties produce fruit more prolifically than wild ones, making them a good choice for some gardeners. In our area there are so many wild ones that actually planting the bushes is superfluous. There seems to be plenty of fruit for humans and birds. Cranberries appeal to robins, crows, sparrows, and game birds. The highbush cranberry attracts game birds and waxwings. Figs usually attract a great variety of songbirds.

Some of these fruiting shrubs, especially the prickly ones, provide excellent shelter and nesting sites as well as food. From the human point of view, they're useful and not much trouble to grow. A little pruning here and there is in order, but insect and disease problems are minor.

Ornamental Shrubs

None of us can grow all of the things we might like to grow, and it may well be that you prefer shrubs that are more decorative than utilitarian. Many of us like to plant at least a few broad-leaved or needle evergreen shrubs, especially as foundation plantings. We've seen how valuable the needle

Robin

evergreens are for shelter to wintering birds, and junipers also produce berries that waxwings and robins eat. You probably wouldn't have to be persuaded to plant azaleas in any case, but isn't it a nice bonus that hummingbirds like the nectar of their flowers?

Cotoneaster is another widely grown ornamental shrub whose fruit is enjoyed by bluebirds, robins, waxwings, finches, and mockingbirds. Catbirds and cardinals join the bluebirds in enthusiasm for both autumn and cherry elaeagnus. Euonymus in its multitudinous forms will also please bluebirds, as well as warblers, sparrows, and mockingbirds. Nataplum's outdoor range is restricted, but in southern Florida the bulbuls flock to it. Hollies have decorative fruit that many favorite birds find edible, but don't forget you need both male and female plants to produce fruit, borne only by the female holly. About fifteen varieties are available.

Both Japanese and European barberry provide bird food, and so do cabbage palmetto, nandina, pyracantha, privet, multiflora rose, serviceberry, viburnum, spicebush, and bayberry. Do be careful of the multiflora rose as an ornamental, though. It's out to take over the world, and pruning it is difficult because the thorns are vicious.

As you can see, it may be more of a challenge to plant shrubs that *don't* provide bird food than to plant those that do. The choice is certainly large. I wouldn't dream of limiting myself only to those things the birds enjoy, but it's easy enough to include some of them and still have room for rhododendrons, mountain laurel, lilacs, flowering almond, and the like. At the very least, all shrubs provide perching sites for the birds. Some of those that don't produce fruit or seeds that songbirds eat attract hummingbirds to their flowers — Japanese quince and rose of Sharon come to mind.

Vines

There are many more plants that bring birds to the yard. Consider the vines, such as English ivy, bittersweet (which can be invasive), and trumpet vine. When weather permits, we like to eat our meals on the side porch. The trumpet vine that climbs up at one corner starts to bloom in July. For four to six weeks it blossoms with abandon, and the hummingbirds visit it often, not the least dismayed that we're dining, too, not six feet away.

If you want something better behaved and less flamboyant, try Boston ivy or grapes. There we run into competition with the birds again. We have no intention of going to the trouble of erecting a support for grape vines, cultivating them,

Shrubs to Attract Birds

The single best way of enticing a variety of nesting birds is to include shrubby hedgerows in the landscape design. In addition to providing shelter, they can offer various kinds of natural food. Common honeysuckle and Hansen's bush cherries are prime examples as are currants, gooseberries, and wild or cultivated blueberries, if you're willing to share.

Honeysuckle

Don't ignore honeysuckle because it's "common." No shrub is easier to grow; in fact, it can become invasive. There are many different kinds of honeysuckle. You can whack it off to suit the space available, and you'll probably have to do just that; but you can let it grow however it wants to in a wild corner. We have enormous clumps of it, bird planted, out on the brink of our marshy area. We also have it planted by the driveway. I'd hate to be without it, considering the way birds flock to it. Common or no, the robins, catbirds, and waxwings don't care.

Corn this high is bird-proof.

and coddling them, without some personal return like jelly and snacks. Let the birds have the wild grapes, and protect the cultivated ones the best you can. If there are enough wild ones available and your domestic ones are close to the house, you won't have much trouble.

Vegetables

Moving to the vegetable garden, you'll discover that the blackbirds are willing to eat your peas if they discover them. Goldfinches like catnip seeds. I've chased robins away from the first ripening tomatoes. (Birds all seem quite attracted to *red* fruit.) Mockingbirds, waxwings, bluebirds, and pheasants love asparagus seeds. It's possible — in my garden at least — to see a variety of birds feasting on dandelion seeds: buntings, goldfinches, sparrows, and nuthatches. Dandelions farther afield are likely to attract game birds.

Depending on how much room you have, you may want to devote a portion of your vegetable garden to bird food. Practically all birds like corn, whether sweet or field corn. Some of them, unfortunately for us, like sprouted corn. Gardeners may have to protect the emerging plants until their root systems have developed enough to anchor them firmly. That gives you a chance to make an imaginative scarecrow. Aim for imaginative, since it's unlikely to be effective.

Millet is a grass-type crop that is very valuable as a bird food. You'll remember that its seeds are the base of many

Plant in Long Rows

If you decide to plant bird-food crops, it's preferable to place them in several long rows, producing more edge. The edge effect is important in attracting birds of various kinds. Millet and sunflowers are two favorite bird foods you can grow easily.

commercially available bird food mixtures and that there are many varieties. You can grow your own sunflowers and let the birds enjoy the seeds after you've had the pleasure of seeing the flowers. Rape, oats, and wheat are other possibilities. For most of us, however, these are not much more than amusing projects to be tried on a small scale. It takes a lot of land and work to grow enough of any of them to be significant in a feeding program.

In reading about plants that birds enjoy in climates kinder than the one in which I garden, I've come across the blue gum eucalyptus, which is native to Australia but will grow (and quickly, too) in warm, moist parts of the United States. It belongs to the myrtle family, and its blossoms are filled with nectar. Waxwings eat the flowers avidly.

Brush Pile

That leaves only the brush pile. If you have a small yard, you'll probably want to dispense with that idea for aesthetic reasons, except possibly as a sometime thing. Large properties can devote some out-of-the-way corner to a brush pile. It will provide welcome shelter, especially in winter.

Ground Covers and Lawns

Some common ground covers provide food for birds. Clover and dandelions in your lawn or a pasture are likely to attract game birds, as will vetch used as a ground cover. Starlings, grackles, and robins will search for insects in any lawn, and most of those they find are ones we'd just as soon be rid of. An expanse of lawn sets off gardens nicely and also provides a measure of safety for the birds at your feeder or birdbath because predators can't easily hide.

A brush pile provides shelter and nest sites.

The poison ivy vine produces a white berry edible by birds. That's one reason the plant is hard to control sometimes. It's often bird-sown.

In any size lot, if you have an area set aside as a vegetable garden, you might construct a small brush pile for the use of birds during the harsh months. A few branches, debris from the garden, and the discarded Christmas tree will be welcome as shelter for winter feeder guests.

It's pleasant that many of the qualities that appeal to us in our gardens, such as color and variety, appeal to birds, too. What's maybe even more reassuring is that a measure of untidiness is welcome to them. If flowers or herbs go to seed, you can bet that some bird will enjoy them. When you don't get around to deadheading the ornamentals or drying the herbs, you can always announce loudly that you wouldn't dream of depriving the birds of a food source.

PART II

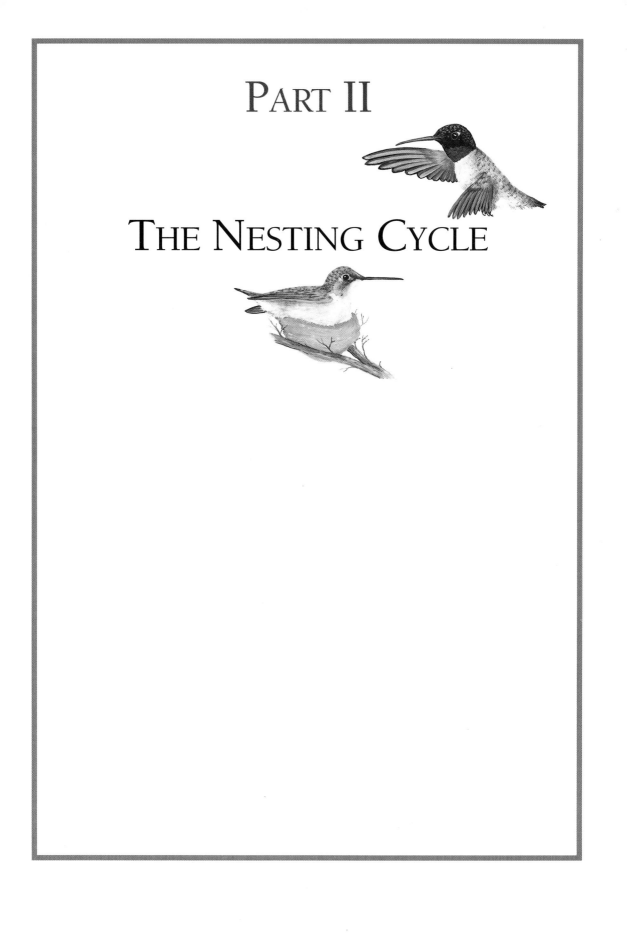

THE NESTING CYCLE

-7-

Sharing the Earth

Mallards

THEIR CRY SENDS ME TEARING OUTSIDE to scan the sky. No matter that the day is cold, that the wind chills me even before I close the door behind me, that the grimy slush in the driveway soaks my sneakers. The first scraggly skein of Canada geese passing northward overhead means that spring is on the way.

The seasonal migration of birds back north signals the visible beginning of their annual nesting cycle. Hormonal changes occurring during the winter eventually send them winging toward summer ranges, there to claim territory, court a mate, and produce offspring. Those biological imperatives attended to, and further hormonal changes ending the reproductive urge, survivors young and old prepare for the return flight to winter quarters.

In the chapters that follow, I'll concentrate on the nesting cycles of birds found in and around gardens north of Mexico and south of the subarctic regions. Whether your home is urban, suburban, or rural, you can encourage nesting birds to take up residence. As discussed previously, food, water, and a sheltered nesting site are the essentials: a feeder, a birdbath, and a birdhouse. Gardeners can introduce plants that produce food as well as shelter. Fortuitous geographical features such as a pond or brook, or your own installation of a pool or a fountain, will increase the number of birds that find your garden attractive. Some birds will accept your nesting boxes; even more will accept the nesting materials you provide. You can observe birds in all stages of the cycle, provided you are willing to be patient.

BIRD WATCHING

Up until this century, field ornithology consisted primarily of exploration and classification. Classification usually meant the actual *collection* of nests, eggs, and even birds. It was an effective way to learn about birds, but hardly beneficial to the birds themselves. The modern technique of identifying birds by sight, using their size, shape, flight pattern, and so on as clues, makes a great deal more sense than shooting them does. However, many keen birders collected all three types of specimens, often many sets of the same species. One source reports an individual who had 235 sets of robin's eggs — surely a case of more enthusiasm than good judgment.

Through the conservation movement that began to attract attention around the turn of the century, awareness of the ever-increasing destruction of wildlife made collection unconscionable to birders and resulted in the activity known as *listing,* in which people keep written records of what they encounter rather than collect specimens. Today it is against federal law to take nests, eggs, or even the stray feathers of birds if you are not authorized to do so.

Eventually the practice of listing led birders to begin reporting nesting data to the Laboratory of Ornithology at Cornell University, the clearinghouse for such information. Careful, documented observation by amateurs has traditionally been highly useful to ornithologists, and today remains a valuable resource.

BIRDS AND HUMANS

Bird numbers fluctuate because of numerous threats to the activities of birds. Human threats have included sport hunting as well as the commercial slaughter of wild birds for meat and feathers; the introduction of bird species that compete with native species; poisonings from lead, pesticides, and other chemicals; and oil spills (auks, for example, will dive when they see one and then surface in its middle).

Yellow-headed blackbirds

Environmental changes also affect bird distribution and population. Lowering of the water table tends to displace birds such as the shoveler, the American coot, the common snipe, and the yellow-headed blackbird. Upland sandpipers are disturbed when land is plowed; they prefer grasslands. Although the creation of artificial lakes, especially large ones, will attract water-loving birds, strip-mining is apt to displace birds, even where there is partial reclamation (and often there is none). Stream channelization and asphalt sprawl displace birds, although soil conservation programs and the protection of woodlands in state and national parks and forests help to maintain existing populations.

Structures such as television towers, skyscrapers, and even windows take an annual toll. Recreational activities like swimming, sunning, and riding vehicles in nesting and feeding areas drive birds away. The draining of wetlands and competition from fisheries endanger the food sources of some birds.

Anna's hummingbird

Bird Counts

In the past, birds were described as being **common, uncommon, occasional,** or **rare** in specified habitats. Such terms were subjective, and their meaning varied according to the experience of the birder reporting or receiving the information. The current practice of *counts* is more appropriate because it provides concrete data less likely to mislead through errors of definition and the vagaries of interpretation. To make these counts, armies of amateur and professional birders gather at a certain time (the Christmas Bird Count attracts a lot of attention) or in a designated area (a flyway, a roosting area, a designated patch of farmland, or near a feeder) and actually count the birds present by species. From these counts, estimates of populations are made.

The terms now used refer to actual numbers counted. **Abundant** describes birds observed in daily counts of up to fifty and season counts of two hundred and fifty or more. **Common** birds are observed in daily counts of six to fifty and season counts of up to two hundred and fifty. **Uncommon** birds are observed in daily counts of one to five and season counts of five to twenty-five; **rare** birds, five in a season. **Casual** means up to three birds in ten years, and **accidental** is up to three in a lifetime. Approximately eight hundred birds can be identified in North America.

There is speculation that human-induced climate changes alter migration patterns. Certain other weather-related problems, not attributable to people, include the unreliability of the food supply, the effect of temperature and precipitation on breeding, the disturbance of flight by heavy winds, and the destruction of birds by storms.

Although many human activities have been injurious to birds, it's worth mentioning that certain birds cause problems for human beings, who are also part of the ecological scheme. Mute swans may eat, trample, or foul the grass in grazing pastures, and Canada geese make serious inroads on grain crops in some areas. Starlings wreak havoc in orchards. Many farmers in England consider the wood pigeon a major agricultural pest and the bullfinch a horticultural pest. Linnets attack strawberries and other soft fruits; tits damage roses; and jays savage the beans. In North America, fecal deposits under the winter roosts of such birds as starlings

Mountain bluebird

Classification of Birds

Like all other living organisms, birds are classified in ever-narrowing categories. They belong to the animal **kingdom,** the vertebrate **phylum,** the **class** Aves, or birds. There are more birds — over 8,500 living species — than any other vertebrate except fishes. The mountain bluebird is an example of a **species,** and *Sialia currucoides* is its specific name; only a **subspecies** is lower on the ladder of classification. The next larger category in which the mountain bluebird is classified, along with other bluebirds, is their **genus,** *Sialia.* Their **family** is Muscicapidae and their **order** Passeriformes, or perching birds.

Altogether there are twenty-eight orders of birds. Orders are usually worldwide in distribution, but families and groups of genera may be limited to a continent or a zoogeographic region (a large area that has characteristics suitable for them), and genera and species may occur only in certain parts of a continent. An astonishing 85 percent of known species and subspecies of birds are found only in tropical areas, and two-thirds of *them* are found only in the humid parts of the tropics. Considering this, it's not surprising that thirty-two of the forty-four families of birds that exist only in the New World are found exclusively in the tropics. A dozen of the New World families can be found in the United States. They are migratory and winter in the tropics. Moreover, they prefer the humid, eastern part of the United States.

and red-winged blackbirds can kill trees. (As many as twenty or thirty *million* red-winged blackbirds may inhabit a single roost in the southern states.) The herring gull can even damage airplanes, shattering windshields and jamming turbine blades. At the same time, of course, birds provide such invaluable services as eating insects and distributing seeds and pollen — not to mention the aesthetic pleasures they offer.

Herring gull

Most birds are specialized, but some are highly adaptable, demonstrating an ability to benefit from human activities as we benefit from theirs. Starlings, house sparrows, and pigeons, introduced to North America by human beings, have adapted so well that many people now consider them nuisances. Some highly attractive native species have also proved adaptable. Peregrine falcons nest in cities to feed on the introduced pigeons. Robins, swallows, barn owls, crows, and ospreys have all profited from the presence of humans, who have provided structures that these birds use and have cleared land, thus making available more habitats of the sort these species require. Certain other machine-made changes to the environment, such as reservoirs and gravel pits, have been helpful to species common to similar natural features. And many species make good use of synthetic nests and nesting materials provided by us.

Does it all balance out? That's a question we must address, if only for selfish reasons. Quite aside from all the other reasons for an interest in birds, there's the old canary-in-the-mine business. Just as miners gauged their safety by the condition of their caged birds, so might we. As the authors of *Kentucky Birds: A Finding Guide* so aptly put it: "In the collapse of certain indicator species we anticipate our own peril." When birds are endangered, humans may be in trouble, too.

-8-

Territory

Bald eagle

TALKING ABOUT TERRITORY in reference to birds gets confusing unless we define some terms at the outset. First of all, habitat and range should be distinguished from territory. **Habitat** is the kind of landscape a particular species needs. **Range** is that wide geographical area in which a species can be found. Breeding range includes any place where nests of the species are found. Wintering range refers to any area in which the species is found in winter.

There are two kinds of **territory.** One is the actual amount of land a pair of birds needs to assure themselves and their offspring an adequate food supply. This amount varies widely. For example, the density of breeding red-eyed vireos may vary from ten territorial males in one hundred acres of open slash land to one hundred territorial males in one hundred acres of virgin hardwood deciduous forest. There must be enough food to feed both the parents and the young. A wren family might subsist in a single garden, but an eagle needs a square mile of land and a kingfisher a linear mile of open water. The size of the territory depends not only on the food supply in a particular area but also on the duties of the nest. Incubating the eggs and caring for the young occupy so much time that the birds must minimize the area in which they search for food.

The other kind of territory is that which the male defends against other males of his own species. Studies of the red-eyed vireo have shown that this, too, varies, from 0.3

Gray catbirds fight over territory.

acre to 2.4 acres, again depending apparently on the relative abundance of food within the territory. Many males defend only a smallish territory in the immediate vicinity of the nest, and the birds use communal feeding grounds. **Colonial** birds (birds that nest in colonies) such as swallows and gulls defend only the territory they can reach while sitting on the nest. The purple martin defends only the nest cavity itself, and the chimney swift defends no territory whatever.

Selection of a territory assures the proper distribution of a species. Familiarity with a given area enables the bird to be self-sufficient and to reserve for its own use the specialized

Cattle Egrets

Cattle egrets have expanded their range dramatically since they first appeared in the southern United States a couple of decades ago; they can now be found as breeding birds as far north as Vermont and New Hampshire and as far west as Minnesota. These colonial birds are often found in the company of herons, though they are less aquatic. They defend a territory of several yards around their nests until incubation begins, at which time they restrict their defense to the nest itself. However, each bird defends a feeding territory as well, often around a particular domestic animal.

A rock dove defends the vicinity of its nest.

nest site it requires, as well as to avoid predators and to find food and nesting material more easily. This efficiency is necessary if the birds are to concentrate on the job of raising their young.

Using a restricted area also helps establish a sexual bond between the parents. Generally speaking, the territory is selected by the male. In those cases where mating occurs before migration, the pair may choose the nesting territory together. Because the male defends his territory, allowing no other males within it, the sexual bond between male and female is necessarily strengthened once the pairing has been accomplished. Other birds of their species are excluded, and . . . out of sight, out of mind.

Having selected a territory, birds defend it against others of their own species. Generally, it is the males who assume much of the defense. One of the common ways of defending territory is by singing. By moving from perch to perch around the perimeter of his territory, the bird signals to others of his species that his turf is off-limits to them.

Territorial behavior varies widely from species to species, so it's important to look at the particulars of individual species (see Part III).

-9-

Courtship

Red-winged blackbirds

MY DEAREST FRIEND has a family of house sparrows living in a birdhouse attached to a tree near her kitchen window. House sparrows are not exclusively cavity nesters; had more luxurious quarters not been available, they would have been entirely happy to nest in the wisteria vine beneath the window. They were not the species she had hoped to attract, but once they laid claim to the property, Winnie didn't discourage them. Nothing is busier than a busy house sparrow, and that goes far in removing any stigma attached to interest in their behavior. Yes, they're as common as dirt; still, their homebody attitudes endear them to all of us devoted to family life, whether in practice or theory. House sparrows are among the few birds we observe that use their nests year round as residences. Moreover, they have persistent family arrangements, are sedentary rather than migratory, and tolerate — encourage is more like it — close observation on a daily basis.

Because of their proximity, Winnie is privy to the considerable domestic activity of the sparrows. They nested first in mid-April, and by July they were raising a second brood and entertaining an entire family of humans with what seemed to be inordinate quantities of generalized bickering and squabbling. Since not all of us have the leisure or the inclination to spend time in blinds observing the activities of more exotic

Cedar waxwings court by feeding each other.

birds, it's pleasant to be able to watch birds close to home. Winnie's observations of the inelegant sparrows enlarged her interest in birds generally.

PROPAGATING THE SPECIES

California quail

Much of what we know about reproduction in birds stems from the pioneering work of Professor William Rowan at the University of Alberta in the 1930s. He studied **photo-periodism** (the effects of differing periods of light) in birds. Working with crows and juncos, he discovered that artificially increasing the length of periods of light in fall and winter caused the birds' sex organs to emerge from the resting state and enlarge. However, this information tells us only about reproductive behavior in birds of the temperate zone, where day length varies with the season. Tropical birds may breed at any time of the year. In Africa and Australia rainfall patterns affect nesting, whatever the season.

The only constant seems to be that wherever the birds, there is always a cycle of rest, which may last several months. During this "refractory" period, the reproductive organs don't respond to any stimulus. Following it, however, birds can become sexually responsive to factors other than light. Warmth, rainfall, food, and the presence of mates and nest sites also stimulate the sex organs. Cold, aridity, food scarcity, a lack of nest sites, and the absence of social stimulation all retard their enlargement.

Differences between the Sexes

Males are ready to mate before females, and they proceed to engage in courtship displays. Their earlier readiness may explain the pronounced **dimorphism** (difference between the sexes) in some birds. This dimorphism may be seen both in plumage (such as that of grosbeaks) and in behaviors. Sage grouse, for example, gather in arenas every morning for a period of weeks. While the females watch from the sidelines, the males fight until a hierarchy is formed. The dominant male eventually mates with most of the females, though males lower on the scale sometimes mate with females while the dominant bird is occupied.

Differences in plumage in males and females of the same species include patterns on wings or tail and the colors of the feet or bill. Brilliantly colored songbirds engage in fewer extravagant courtship displays than do the less colorful; it's thought that the bright plumage is sufficient to attract the female. (There are exceptions, of course, such as the peacock, which engages

Who's Who?

North American laughing gulls have a peculiar problem, though not a unique one. Because the sexes look alike, a male charges another gull to find out its sex. Another male invariably will countercharge. If the charged gull does not return the action, the male turns and faces in the opposite direction. Both gulls turn their heads from side to side, and then the female initiates a symbolic food-begging ritual. Eventually, she crouches to indicate receptivity to copulation.

Rose-breasted grosbeak

Imprinting

In *The Courtship of Birds,* Hilda Simon reports a peacock that displayed only to giant tortoises, with which it grew up, and a jackdaw that offered food only to the naturalist who raised it. Both birds were **imprinted,** a phenomenon familiar to anyone who has raised a newly hatched gosling where no adult geese were present — or an infant mammal, for that matter. The babies assume that the first thing they see is a parent. Unless the impression is corrected by association with others of their species, they persist in the faulty conclusion and demonstrate abnormal behaviors.

Ring-necked pheasants

Mallards

Black-chinned hummingbirds

Scarlet tanagers

Yellow-headed blackbirds

In these and other dimorphic species, males and females differ not only in plumage but also in behavior.

in elaborate displays despite his brilliant plumage.) Spring plumage is nuptial plumage, whether the bird is breeding or not. Although the prenuptial molt may be only a partial one, the eventual breaking off of the neutral feather tips of winter plumage reveals the brighter colors. (You may have noticed how speckled the starlings look throughout the winter and how glossy black they become in spring. The speckles are merely colors at the tips of the feather.)

An interesting feature of plumage is that in species in which the sexes look alike, the male and female usually share equally in nest building, incubation, and care of the young. Barn and tree swallows are examples. Wrens, chickadees — even the raucous jays — share domestic chores. On the other hand, where is the male red-winged blackbird, rufous hummingbird, or pheasant when his mate is busy with nest duties? For that matter, what rooster would be caught dead in a nest?

Display

Courtship displays in male birds include strange postures, loud vocalizations, and manipulation of sacs on the throat or feathers of the crest, ruff, or tail. Visual displays can resemble dances, and some involve what ornithologists call nonvocal songs — drumming, or the whirring of wings. (In addition to those displays intended to attract females are defensive displays, threats against other males of the species in defending territory; and distraction behaviors, simulating injury or sickness, by which males and females try to lure enemies away from the nest.)

Displays enable the sexes to attract each other even though they're rivals for food. The male's sexual displays attract females and repulse other males. These displays continue after copulation to maintain the pair bonds that ensure proper care of the young. Some sexual displays, such as male nest building or courtship feeding, are symbolic. The begging display — usually the gaping and the pecking of juveniles — is sometimes also a courtship display; sexual pursuit often accompanies it. Aquatic birds pursue on water; others pursue in the air or on the ground. A few birds, such as hummingbirds and birds of prey, court exclusively on the wing. During the period of active sexual display, males and some females of certain species, like cardinals and robins, become

Offerings of Food

In the courtship of waxwings, males offer berries to females in much the same manner and spirit that people might offer chocolates. Food offerings help species in which males and females look alike to distinguish the sexes. Food accepted from a male tern in the properly submissive way indicates he has found a female; if his accompanying pecks provoke a beak-grappling contest, he's guessed incorrectly.

A Courtly Dance

Whooping cranes share the responsibilities of bringing up their young and share also in their courtship display. The male approaches the female and bows. After that, both birds display, with marvelously synchronized flaps, steps, and stances, at the end of which the male may leap completely over the female. Her opinion of this bizarre demonstration is apparently favorable.

A male painted bunting struts and spreads his tail to attract a female.

so aggressive that they fight their own reflections in water or in a hubcap. Painted buntings are so pugnacious toward other males that sometimes their fights are fatal.

THE BREEDING CYCLE

Trumpeter swans, which mate for life, pair even before they are sexually mature. During courtship they engage in a mutual display in the water. Facing each other, with wings half spread, they rise out of the water and then return to it, swimming in circles.

The pairing period usually starts at the end of the spring migration. (Some large birds that don't breed until they're two or three years old spend all of their first year in their winter quarters.) Generally, the male begins to demonstrate his range of courtship displays as soon as a female arrives on an established territory. The female is usually unreceptive to copulation at first; the displays eventually stimulate her to accept it.

Once they have chosen one another, the pair builds the nest, mating occurs, eggs are laid and incubated. The eggs hatch, the young are raised, and the family breaks up. Breeding cycles are timed so that the fledglings leave the nest when the climate is most favorable and food most plentiful. In the northeastern United States, for example, returning birds rest and then lay eggs in May, but wanderers like cedar waxwings and permanent residents like American goldfinches lay in summer and feed their young on berries and weed seeds, not insects. Since they're in no hurry to get anywhere on schedule, they can afford an apparently leisurely approach.

Polygamy and polyandry are more common to game birds than to songbirds, but there are exceptions. Male red-winged

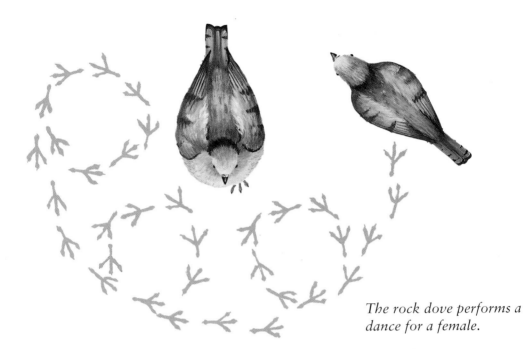

The rock dove performs a dance for a female.

blackbirds average three mates each. Common grackle males also leave the females incubating the eggs and go off to seek another mate.

Most birds, however, are monogamous for at least one nesting season, and nonmigratory birds are likely to remain mated for a year or longer. Birds will usually mate again if the partner dies; mortality is high during the nesting season, but there is always a floating population of unmated birds. When one sex greatly outnumbers another, homosexual pairing sometimes occurs, especially among captive birds. Some few species change mates between broods. In crowded colonies of red-winged blackbirds, a male may find it difficult to defend his territory against the great numbers of other males. Sometimes a female red-wing mates with more than one male.

-10-

Birdsong

Both male and female cardinals sing during the breeding season.

BIRDSONG IS PART of the courtship displays. Like courtship displays such as drumming or the whirring of wings, vocal song requires both specialized organs and a certain development of the brain. Song demands more learning (including imitation) than do the calls of birds. Calls are often a single sound, such as squawking, peeping, or screaming. Even if the call is repeated, it's likely to lack variation. Calls serve a variety of purposes, however, including scolding, warning, begging, summoning, or indicating a location.

Songs consist of a series of notes in a somewhat discernible pattern. They tend to be more leisurely, more relaxed, and they first occur as youthful play. They are relatively short because of the limited attention spans of birds.

Females don't usually sing as much as males do, but they're more likely to sing in the tropics than in temperate zones. A female chickadee has only the *fee-bee* song; on the other hand, female cardinals sing as well as their mates. Singing in males is closely correlated with the presence of male hormones and is at its loudest during the breeding season.

Although birdsong is part of the courtship performance, it's used for more than announcing a claim to territory or attracting a mate. Through their songs birds can identify not only their own species but particular individuals, as well. That can be very important. Birds may produce, say, twenty eggs in a lifetime and only just manage to replace themselves, which is surely indicative of the perilous lives they lead. They must be able to tell who's who, as well as to recognize the calls that signal distress, hunger, and alarm. Visual signals aren't enough. Birds of the same species aren't just similar to one another; many females of *different* species strongly resemble each other. Examples include multitudes of sparrows, purple and house finches, dickcissels, siskins, pipits, and longspurs. No one has yet decided whether birdsong is music or language, poetry or prose, but it must be distinctive at a distance for it to serve its purposes.

Imitation is common among songbirds, and we suspect that imitation signifies interest in sounds as such. Birds in a given region develop what might be called dialects, just as people do. Bird dialects always have this geographical basis despite a hereditary predisposition of certain basic types of utterance. Isolation of a group of, say, chickadees, from others of their species, promotes dialect formation. Call notes are more constant than songs because less imitation (learning) is involved. "Accents" show up better in the learned songs than in innate calls.

In her book *Bird Songs*, Norma Stillwell tells how she and her husband took to the road and began taping birdsongs as a more or less full-time activity after they retired. During travels from their base in Texas, they discovered a wide variation in song, depending on where the birds lived. Gradually

Many Reasons to Sing

Though prenuptial singing is the richest, songs accompany nest building, incubation, tending the young, the departure of the young, roosting, autumn, migration, and winter. Through their vocalizations, birds proclaim territory, issue mating invitations and invitations to the nest, request relief from the nest, maintain and strengthen a sexual bond, and abet flock formation.

Dickcissel

Singing for Joy

Some students of birdsong believe that though a correlation between need and skill exists, the skillful sing even if the need is not pressing, and with more frequency and variation than seems required. They point out that just as human music has its purposes — such as stimulation for war, love, and magic — so, too, birds sing to express feelings. No one would argue that birds are as musically sophisticated as human beings are, but however primitive their song is, it persists when no actual need can be demonstrated.

Some experts believe birdsong is sometimes used as an emotional release, described either as an excess of energy or an outpouring of sheer joy and called *ecstasy song*. Such song is produced randomly and is sometimes accompanied by the so-called *ecstasy flight*. Believing that this activity is motivated by a hope of harmony and escape from discord and boredom will not be easy for everyone to do, but I'm willing to entertain the idea.

Secondary Song

Sometimes birdsong is divided into primary and secondary song. Primary song includes: male territorial song to repel other males and attract females; signals used to coordinate activities, especially with those of the mate; emotional song (reason unknown); and female song, which is less common in North America than in other parts of the world. Secondary song includes whisper song, which is territorial song without territorial implications (that is, it's not intended to exclude rival birds or even be heard by them) and is produced by either sex at any time; and subsong, which is unlike territorial song and is produced mostly by juveniles. Subsong may be compared to the play of young animals; it's a kind of practicing by the immature birds.

Song sparrow singing.

they got more and more sophisticated equipment, learning the best techniques for using it as they went along, and continued traveling during the "song season," going south in March and following the birds until about early July. They recorded three volumes of *Bird Songs of Dooryard, Field and Forest,* which have been widely distributed.

Some of the songs that birds create are unique to the individual, even though the song developed initially through mimicry. We're all aware of the ability of mockingbirds to mimic birdsongs as well as other sounds. It's actually possible to learn the songs of other species by listening to mockingbird imitations. Other members of the *Mimidae* family may sound much like the mockingbird, but they can identify their own species fairly easily. Catbirds sing a phrase only once; thrashers repeat it, and mockingbirds repeat it several times. It's estimated that a thrasher is capable of uttering a thousand recognizably different songs. A number of birds can even be taught to "talk," although this is possible only with captive birds — they apparently will not do it in the wild. Maybe captives learn to talk as a form of social adaptation — to get the attention they need.

The songs humans enjoy most tend to be the songs of otherwise inconspicuous birds — precisely the birds that have the greatest need of distinctive songs, since they blend into their surroundings so well that they're hard to detect even by others of their own species. Each species usually has two or more distinct types of song, which may have regional varia-

Wood thrush

tions. Cardinals have two dozen or more, and Aretas Saunders noted 884 variations in song sparrow songs.

The best singers are perching birds, especially the thrushes (Muscicapidae), mimic thrushes (Mimidae), and wrens (Troglodytidae). Families, of course, differ ecologically and presumably in their ancestry — and that correlates with the degree of their song development. (If song helps assure success in breeding, through evolution birds eventually develop the necessary anatomical equipment to produce song.)

Moreover, aesthetic ratings of the songs by human beings correlate with the physical development of the birds. We rate the Mimidae — mockingbirds, catbirds, thrashers — highly as singers, and all of them have seven pairs of vocal cord muscles, whereas most other birds have a single pair.

Though the characteristics of song are influenced by the construction of the syrinx, or vocal cord, perfected song is probably learned. Thus, the evolution of song marches right along with the evolution of the birds, from lowest to highest in development. Gulls and birds of prey, low on the evolutionary scale of birds, have almost no song, but perching birds such as thrushes, in the highest order of birds, sing a lot. Small size also tends to produce song, perhaps because songs make the singer recognizable from a distance.

Song itself is characterized by pitch, intensity, volume, and quality. Various songs are distinct by reason of timing, repetitiveness, notes, trills, and phrases. Edward A. Armstrong, a respected authority on birdsong, display, and behavior, believes that two or more phrases in a regular succession may correspond to a sentence. He believes that songs communicate the location of food, water, nest sites, roosts, and predators. Moreover, they may convey information such as species, sex, individual identity, status, sexual motivation, need, aggressiveness, alarm, or fear, and the location of an individual. It seems safe to assume, at least, that song gives clues to the size of the territory, the number of males in an area (since males sing more in dense population areas), and the phase of the breeding cycle.

Variations in birdsong seem endless. The "whisper song" may be heard during bad weather, including periods of extreme heat, but some birds sing quietly on the nest regardless of weather. Swainson's thrushes sing the whisper song during migration, and vireos when first arriving in their territories. Blue jays sometimes sing it when they think no one is around — it wouldn't do to ruin their image as screamers.

When defending their territory, birds sing most often from an open perch, usually at a distance from the nest, although some males sing while actually on the nest. Singing also

Time of Day

Birdsong is more common at dawn and dusk than at any other time, but you'll hear less of it during cold snaps and in periods of wind or heavy rain. Some diurnal birds — mockingbirds, nightingales, and the marsh wren — sing at night. Birds start singing when they're young; they have their full repertoire of songs by the time they raise their first brood.

Blue jay

Yellow warbler

occurs in flight. The loudness and frequency of territorial songs decline as the nesting season progresses. Although females sing less frequently than males, duets of pairs may reinforce pair bonding. Since songs are most common among birds living in dense habitat, it may help them communicate when visual contact is difficult.

A lot of this free-floating melody is leading up to something: having proclaimed a territory and attracted a mate, the male is about to become a parent. Whether he's helpful in the next phase — nest building — depends on his species.

Studying Bird Song

In 1650 Athanasium Kircher published a book that described some birdsongs in musical notation, but what is considered scientific study of birdsong didn't get underway until the twentieth century. Eliot Howard, an English ornithologist, is credited with laying the foundations for such study at the beginning of the century.

In 1904, F. Schuyler Mathews published his *Field Book of Wild Birds and Their Music,* a famous source that uses normal musical scales to illustrate birdsong. (It has long been out of print, but many libraries have copies.) It is very complete — though a little weird in its subjectivity. In his preface Mathews says, "He [the bird] can not sustain a melody of any considerable length, nor can he conform to our conventional ideas of meter, but he can keep time perfectly, and a knowledge of his rhythmic methods is, I believe, the strongest factor in his identification by the ear!"

In 1935 Aretas A. Saunders published *Guide to Bird Songs,* which introduced various formulas for identification and phonetic syllabification; it is still considered helpful. The development of electronic sound recording and spectrographs has made possible an objective study of birdsong, which is part of a whole complex of integrated behavior. Even though individual songs may vary widely, characteristics are usually present that make possible the identification of the species.

-11-

Nesting

Barn swallow nest

THINK OF NESTS AS CRADLES for eggs and nurseries for young birds. Just as the world still has relatively primitive birds such as loons in addition to the entire range of more highly developed birds, so, too, bird nest development extends from bare gravel to the marvelous hanging nests of Africa's weaverbirds.

Birds are descendants of reptiles and began emerging in the orders we know today during the Paleocene and Eocene eras, the earliest part of the Age of Mammals. Nest building probably developed as birds became warm-blooded and could no longer leave their eggs to hatch in the heat of the environment. Ideally, nests should protect not only the eggs but also the incubating parent.

The kind of nest a bird builds is directly related to the degree of evolution its species has achieved; it's assumed that birds and nests evolved together. The first nests, at the beginning of bird history, were probably depressions in the ground or natural cavities in rocks, tree roots, and hollow trees. Even today some birds merely choose a *place* but don't modify it in any way — the female lays the eggs and, evidently, hopes for the best. Whip-poor-wills are one example: the female merely places her two eggs on a bed of oak leaves on the ground. The killdeer lays her eggs on a bed of wood chips, grass, gravel, or cinders and on the graveled roofs of buildings, a location also chosen by common nighthawks and one that seems a great deal more secure than the ground. Most

For more on what specific birds need in the way of habitat and nesting materials, *see* the chart "Habitat Requirements," starting on page 290.

nests, however, protect the eggs and maybe even the hatchlings.

When some birds began to elevate their nests, it became necessary for them to gather nesting materials, probably twigs and sticks, to serve as a platform. Phoebes and robins use platforms to this day. Eventually certain birds, especially the highly developed passerines, built complicated cup-shaped nests.

CATEGORIES OF NESTING SITES

Nesting sites are as variable as birds themselves, but they can be roughly grouped into four categories. First are those birds that use the ground, cliffs, or beaches. Among the ground nesters are game birds such as pheasants, quail, grouse, and turkeys. Some small birds such as juncos, sparrows, and

Brown-headed cowbird

Northern oriole

Screech owl

Killdeer

Different nesting styles.

warblers also nest on or very near the ground. So do the hermit thrush and some doves. We can't do very much to help them except to maintain such areas for them, safe from mowing and burning.

Another group, a large one, includes some favorite songbirds such as orioles, robins, cardinals, catbirds, mockingbirds, thrashers, kingbirds, grosbeaks, and other finches. They prefer to nest in trees, shrubs, and vines and can be enticed to our yards if we provide adequate cover for them.

A third group likes an aquatic habitat such as a swamp. Not many of us can provide a swamp, but we can, in our towns, help protect wetlands for them. Apart from the herons, egrets, ducks, and gulls that you'd expect in such habitats, you'll also find the yellowthroat, blackbirds, some sparrows, and boat-tailed grackles.

Birds that will use birdhouses, the cavity-nesting birds, are in the fourth category. These include many familiar garden birds: bluebirds, chickadees, nuthatches, wrens, tree swallows, titmice, flickers, woodpeckers, starlings, and martins. Among the not-so-common species that will use birdhouses are great crested fly-catchers, sparrow hawks, screech owls, and wood ducks.

Artificial Nests

Ornithologists have discovered that artificial nest structures are useful for birds other than cavity nesters. Ospreys, eagles, cormorants, and owls accept platforms; terns, loons, ducks, and geese accept artificial floating platforms. Snail kites, which are an endangered species, accept metal baskets constructed for them and seem to prefer them to building in marsh vegetation.

SPOTTING NESTS

Nests are ordinarily as inconspicuous as possible, a factor in survival. Considering that the typical figure for nesting success (youngsters that achieve maturity) is 20 percent, it's obvious that nesting birds face horrendous problems. The best time to find and identify nests is in the winter. You'll not be disturbing any birds, and the nests are easier to find when leaves are gone from the trees and ground vegetation isn't growing.

You don't have to mount an expedition to look for nests, either. I'm always surprised at the number of nests I can see from my car window. Many birds build near the road, even when it's a busy interstate highway, and many nest in city parks. Wherever you see a stand of mature trees, however small, you'll find nests. If the trees are tall enough, you'll see enormous nests. Nests of birds that prefer brambly sites are harder to spot, even after the leaves have disappeared.

Hiking and horseback riding are good ways to see nests in all their diversity. Some nests are easier to recognize than others — it's difficult to identify the bare scraping of ground that suffices for certain species, although slightly more highly developed birds line such scrapes. Others use natural mounds

Woodcock

American woodcocks, found in damp woods, nest 300 feet from the singing ground (the area above which the male performs his courtship flight song), in an unconcealed depression in dead leaves rimmed with twigs and lined with pine needles.

Hummingbird nest, actual size, with male black-chinned hummingbird.

or floating clumps of vegetation, while ground holes as well as tree holes may be nesting sites. The cup-shaped nest is considered the classic, however, and it can be found on the ground or floating on a raft or wedged in the fork of a tree. It may be domed or otherwise enclosed, suspended, or bound with mud.

Humans and the Nest

As far as observing active nests is concerned, our respect for the serious business of reproduction should monitor our curiosity. There's no denying the pleasure of watching the procedure from the beginning of nest building to the first flight of the fledglings. That, after all, is one of the reasons we put effort into erecting birdhouses on our property or landscaping it to attract birds. Birds that nest close to houses are unlikely to find discreet interest in their activities annoying, but "discreet" doesn't include handling nests, eggs, or the young. If a parent bird arrives on the scene, it's prudent to retire politely.

Too much attention to a nest may attract the attention of nest predators, including our pets. Birds that nest farther from human habitation are likely to resent our presence (red-winged blackbirds will attack snoopers). In addition, people being nearby may prevent a parent bird from returning to a nest to care for the young. A pair of eagles in Norway circled their aerie for days because campers too close for comfort frightened them. The eaglets died from hunger as a result.

DIFFERENT NESTING STYLES

Some birds, like ducks, nest communally (in the same area with other species), some nest in colonies, and some accept nesting boxes. Martins are colonial nesters and are entirely willing to use birdhouses. The aforementioned weaverbird of South Africa builds apartment complexes of up to one hundred units, which belong to the flock. Ani cuckoos in tropical sections of the Americas build their basin-shaped nests as a communal home; several females lay eggs in the nest and the whole community cares for the young, which in turn may stay and help out with the next brood. Except for seabirds, birds usually nest in their feeding territory. The site of the nest is a major clue in the identification of its owner.

Nest Building

Each species makes a distinctive type of nest, and there's considerable variation in the materials used. On the whole, birds use readily available materials, which itself accounts for variation depending on location. Some birds, for protection, nest close to an aggressive species of animal. A few of the cavity nesters also use their nest sites as roosts and shelters.

Who Builds the Nest?

Choosing the site of the nest and building it are tasks dominated by the female. In polygamous species, the female does it all, but monogamous birds often share equally or divide the tasks.

Female orchard oriole selects site for her nest.

Birds that collect food in their claws also collect and transport their building materials the same way; those that use their bills to collect food use them to collect nesting material, as well.

The type of nest used by a species is thought to be intrinsic, but some learning of techniques apparently takes place because first-time builders are less skillful than experienced ones are. Those birds that build new quarters for additional broods spend less time and energy on each successive nest. Nests may take a day or weeks to build, depending on the length of the breeding season. Some birds continue adding to their nests after they're in use. The time spent on building a nest also varies with weather and area; for example, birds in the Arctic spend less time on the chore than birds in temperate climates do.

In North America, northern (Baltimore) orioles, bushtits, and woodpeckers build the most complicated nests. In other parts of the world, there are even more complex bird nests. Birds are highly adaptable and build deeper nests in places subject to strong winds, for example. Some nests are so tiny that they are hidden under the incubating bird; some are so enormous that the bird disappears within them. The records for large nests in North America go to a bald eagle family in Vermillion, Ohio, whose nest was 12 feet deep, 8½ feet in diameter, and weighed two tons, and a pair in Florida using a nest 20 feet deep and 9½ feet across.

Most passerines build a new nest for each brood, but wrens, hawks, eagles, and owls will use the same nest for

Great horned owls will use an abandoned eagle's nest.

years, and bluebirds will use the same nestbox over and over. Certain tree-cavity nesters don't excavate their nest sites but use whatever cavity is available. Some use abandoned woodpecker nest holes; hawks and great horned owls will appropriate an abandoned eagle, hawk, or osprey nest. Occasionally birds use old nests for winter shelter. Robins may use an old nest as a foundation for a new one. Occasionally up to three chimney swifts or three cliff swallows will use the same nest — some birders speculate that a sort of apprenticeship is involved.

NESTING MATERIALS

Nesting materials are usually of plant origin, but mud, animal hair, and feathers are also used. Hummingbirds and parula warblers use spider silk to strengthen their nests. Some birds incorporate cellophane, tissues, yarn, and newspapers into their nests. Ospreys have been known to work bathtowels, garden rakes, old rubber boots, and straw hats into their enormous trash-filled nests. Hawks decorate theirs with green leaves, replaced regularly. Linings, if any, are often soft and warm.

Orioles, which usually use milkweed plants, will take string or twine, grass, cellophane, wood shavings, cotton yarn, and strips of cloth to weave into the construction of their elaborate nests. Robins, vireos, yellow warblers, chipping and song sparrows, goldfinches, and cedar waxwings also will use these materials.

Osprey

Offering Nesting Materials

You can make certain items available for birds to work into their nests — hair clippings, shed dog or cat fur, short pieces of string or yarn. Some people drape the offerings over shrubs, but you can put them in a container nailed to a tree to keep things tidier. Make sure the string and yarn are no longer than six or eight inches because longer pieces can form loops or nooses and strangle the birds you're trying to assist.

Supply birds until August to take care of successive broods and late nesters. The birds may raid your garden for some odds and ends. They've been reported to use yarrow, rue, thyme, and chrysanthemums in nest construction. Note that all of these plants are herbs that have insecticidal properties.

Long String Can Be Dangerous

Sometimes nesting materials can harm the nestlings. Birders are warned not to offer nesting materials such as long pieces of string or yarn that might strangle a bird (4 inches is a safe length). We had a catastrophe in our barn last summer. A barn swallow pair had used horsehair in their nest, and one morning we found a dead fledgling whose feet had become entangled in it hanging from the nest.

-12-

The Egg

For more specific information about eggs and how different birds incubate them, *see* the chart "All About Eggs" starting on page 297.

THOUGH IT'S PROBABLY SUPERFLUOUS to fall into raptures and rhapsodies on the perfection of Nature and the absolute marvel that an egg is, I must admit that I continue to be amazed by it.

An egg consists mainly of food on which a fertilized cell feeds. An unfertilized egg is rare among wild birds. (Although an egg sometimes remains in the nest after the others have hatched, it is more likely to contain a dead embryo than to be infertile.)

The size of the egg depends on the size of the bird laying the egg and the development of the bird at hatching time. An infant hatched with down, ready to follow its parent, necessarily requires more nourishment (hence a larger egg) than one born naked and helpless. The largest bird egg known is that of the ostrich. It's up to nine inches long and six inches in diameter. The shell, which is a quarter-inch thick, would hold a dozen to a dozen and a half chicken eggs. The ruby-throated hummingbird, on the other hand, lays eggs the size of large peas, about a half-inch long. The vervair of Haiti and Jamaica lays eggs half that long.

The shape of the egg varies with family. Seabirds usually lay eggs of pyriform (pear) shape, presumably to prevent them from rolling off a cliff ledge. Owls often lay spherical eggs. Both shorebirds and bobwhites lay pointed eggs — they can fit more into a nest that way.

The color of the eggs seems to be regulated by the bird's need to foil predators. Eggs laid in the open tend to be

Eggs come in a variety of sizes, colors, and shapes.

blotched and have the heaviest pigmentation, whereas cavity nesters usually lay white eggs. Some naturalists think at one time all eggs were white. The color sometimes changes during incubation. The large end of the egg comes down the oviduct first and tends to pick up more pigment from the cellular walls.

The surface texture of the eggs also varies. Water birds tend to lay rather greasy eggs; some have a pitted surface and others, like those of grebes and cormorants, have a chalky surface. Woodpeckers lay glossy eggs. Passerines generally lay lusterless eggs, one per day, in the morning, until the clutch is complete.

THE CLUTCH

A **clutch** is the number of eggs that comprise a single nesting. The word apparently comes from the Scottish word *cleck*, meaning "to hatch." Some authorities link it to the *cluck* of a hen. If the first clutch laid is destroyed, the second clutch is usually smaller. Experiments have shown, however, that some female birds will go to great lengths to make sure there are eggs to incubate. If eggs are removed from a nest, some species continue laying until the "right" number of eggs is in the nest.

A flicker whose eggs were taken as soon as they were laid produced seventy-one eggs in seventy-three days. Such

How to Describe an Egg

In speaking of the color of eggs, certain conventions are observed. A background color is listed and the markings specified as blotched, spotted, dotted, splashed, scrawled, streaked, marbled, wreathed, capped, overlaid, or underlaid. Both **wreathing** (a band of another color) and **capping** (one end of a different color) usually occur at the larger end of the egg.

egg-laying behavior is called **indeterminate**. Some species are called **determinate** layers: the number of eggs they lay seems to be genetically determined, and the female lays exactly the same number of eggs whether any are destroyed or not. The sandpiper is one such bird. In indeterminate layers, the number of eggs laid seems to be determined by geographical location of the nest (whether or not it's a difficult environment in which to survive) and even by the amount of food available.

Species that produce only a few eggs may be species whose offspring have a high survival rate. A kinder environment can affect clutch size — clutches are smaller in the tropics and become larger the farther north a bird breeds. Ducks and gallinaceous birds (chickenlike birds, including quail and grouse) have the largest clutches, from eight to fifteen eggs. Loons, most raptors, and herring gulls, in contrast, usually lay just two eggs a year.

Clutch size also varies from year to year within the same species, depending on available food. Both the number of eggs and the number of clutches seem to correspond to the security of the bird in its normal environment and to its lifespan.

Dump Nests

Some ducks will lay in whatever nest is handy; such nests are called dump nests. As many as forty eggs have been found in the nest of a wood duck — which itself lays only ten to fifteen eggs.

THE INCUBATION PERIOD

Eggs start their development in the oviduct and stop when air hits them. When incubation begins, development resumes; most eggs can probably remain alive without incubation for three or four weeks. The development of the embryo in the egg is rapid. The vascular system develops first, and breathing begins a day or two before hatching, respiration occurring through the porous shell. The egg becomes lighter during incubation because of the evaporation of water through the shell. It also becomes more brittle.

In some species, male birds as well as females incubate eggs; in those species in which they don't incubate, the males often protect or feed the female. (Some females leave the nest to feed, the frequency varying with the species.) A female spotted sandpiper lays four eggs and leaves them for the male to incubate for three weeks while she hunts up another male and lays a second batch. That accomplished, off she goes in search of still another mate. She may lay up to five clutches per season. This behavior, found also among phalaropes and several other shorebirds, is thought to be the result of their high rate of nest predation.

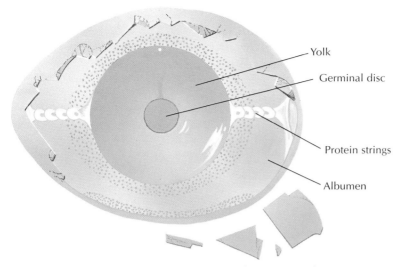

The anatomy of an egg: The spot in the center (the **germinal disk***) grows into a chick, in fertilized eggs. The yolk around it is the chick's food. Anchoring the yolk are* **strings of protein***. The* **albumen***, or egg white, cushions the yolk and protects it from bacteria.*

Some birds develop bare areas on the ventral, or lower, surface of the body, where there is a concentration of blood vessels that provides more warmth for the eggs. This is called the **brood patch,** or **incubation patch.** The male of the species, if he incubates, will develop a brood patch, too. Ducks and geese make their own brood patches by pulling the down from their bodies — which also serves to line the nest and cover the eggs when the parent is absent. In those species in which no brood patch appears, the bird fluffs out its feathers so that the egg is next to the skin.

During incubation, the bird turns the eggs with its bill each time it enters the nest to incubate. The length of incubation varies. As with the size of the egg, it correlates with the size of the bird and the development of the hatchling. Most birds wait until all the eggs are laid before beginning incubation. Those species that begin incubation as soon as the first egg is laid have hatching at different times. Cold weather can delay hatching a day or more.

HATCHING

Noises may be heard from within the shells of some birds a day or so before hatching; interestingly, the noises will cease instantly if the parents sound an alarm note. Hatching may

Penguin Parenting

The prize for the most interesting of all incubating behaviors goes to the male emperor penguin. The female lays her single egg in May and leaves the male to care for it. He places the egg on top of his feet and settles his body over it to keep it warm. Other males cluster with him side by side, incubating their own eggs and helping to keep each other warm in the bitter temperatures. Going without food until the egg hatches, he loses about twenty-five pounds, a third of his body weight, during the two-month incubation period. Meantime, the female has been eating as usual; she returns when the chick hatches and joins the other females, who care for the chicks communally. The male presumably goes to lunch.

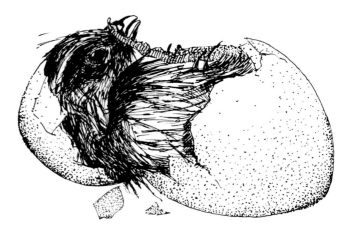

Baby bird coming out of shell.

take anywhere from several hours up to forty-eight hours. Water and game bird eggs usually hatch all at once; passerine eggs, within a single day. As the time of hatching approaches, the parent birds become obviously excited.

In a few species, the parents eat the eggshells after hatching is completed, but most just remove them promptly, presumably because of their strong odor.

There are two types of hatchlings, **precocial** and **altricial.** Precocial birds are covered with down at hatching and can leave the nest the day they hatch. Such birds have a longer incubation period than altricial birds. Altricial birds are naked and blind at birth.

THE CASE OF THE COWBIRD

Many bird lovers despise cowbirds because they deposit their eggs in other birds' nests. Cowbird chicks are usually bigger and more aggressive than their fellow nestlings and thrive at their foster siblings' expense. But like all other members of the animal kingdom except humans, birds are amoral; the concept of morality is not relevant. Cowbirds are not being wicked; they are behaving like cowbirds. Survival, not "virtue," is the major motivation in the lives of birds. The issue is the balance of nature, not moral judgment.

The word *parasite* has acquired connotations of evil, and, therefore, terming the cowbird a "social parasite" conjures up all manner of images of cowbirds playing nasty tricks on less cunning birds. Nonsense. It's merely another way of life, a modification of behavior that has proved beneficial to cowbirds. Though I'm far from a crusader, my interest in defend-

Tools for Hatching

Ordinarily, the parents don't help out in the hatching process. Their help could rupture blood vessels. The baby within has special tools for getting out of the shell. One is the **egg tooth,** used to pip the shell. This is not actually a tooth, of course, but rather a rough, somewhat horny spike on the tip of the upper bill. It falls off a few days after hatching. The infant also has what is called a **hatching muscle** on the back of the head and neck, which gives added strength to the egg tooth. It soon disappears, too.

ing the habits of other inhabitants of the globe has led me to seek out information on the maligned cowbirds.

There are several kinds of cowbirds, in three genera: *Tangavius*, *Agelaiodes*, and *Molothrus*. The members of the genus *Agelaiodes*, which has two species and three races, are nonparasitic. Of *Tangavius*, only *T. aeneus*, the red-eyed cowbird, ventures into the United States, in West Texas and Arizona. It commonly parasitizes orioles.

The third genus, *Molothrus*, is typical of the cowbirds most of us know; they have pointed wings, and the male has darker plumage than the female. They have a courtship display and typical song, and are parasitic. There are three species and ten races.

It's the brown-headed cowbird, or North American cowbird, *M. ater*, that's common in North America. This species uses the nests of 195 other species. These cowbirds migrate and are considered polygamous, the male more so than the female. Normally the female lays five eggs (but one raiser of fledglings reported fourteen eggs), and the incubation period is ten days, which is generally shorter than the incubation period of its host. The brown-headed cowbird always chooses nests in which the host eggs are smaller than its own. Occasionally its eggs are found in an unsuitable nest — that is, one in which the host would feed the nestling inappropriate food.

Bay-winged Cowbird

Note that the bay-winged cowbird is the most primitive of all the cowbirds. Moreover, it has its own parasite, the screaming cowbird, *Molothrus rufo-axillarius;* the young grow up amicably together in the same nest.

Brown-headed cowbird

The cowbird egg is always larger than the host eggs.

No Time to Incubate?

No one is certain how parasitism developed in cowbirds, whether it was an evolutionary process in response to need or the result of a few mutations. One interesting theory of how cowbirds began the practice of depositing eggs in the nests of other birds suggests they couldn't spare the time required for their eggs to incubate in one area since, by trade, they were followers of buffalo (at one time they were commonly called buffalo birds). Some ornithologists think they may not have been able to distinguish between an old and a currently inhabited nest, though most birds can.

It's not true that cowbirds puncture the eggs of other species. It was Audubon's theory that the foster parents of the cowbird neglected incubating their own eggs while trying to feed the hatchling cowbird. Their own eggs, therefore, became rotten and so both hatchling and parents disposed of them. He said the cowbird hatchlings were unlike those of the European cuckoo, which jostle their foster hatchlings out of the nest. Normally the host is able to fledge some of its own brood.

Not all recipients of cowbird eggs actually rear the young. Of the 195 species in whose nests brown-headed cowbird eggs have been known to be laid, ninety-one have been reported as actual foster parents. Red-eyed vireos sometimes recognize the eggs as alien and remove them from the nest; the yellow warbler always recognizes them and builds another nest on top of one containing a cowbird egg (repeating the procedure as many as eight times to escape the offenders). It's not uncommon to find more than one cowbird egg in a single nest — a mockingbird nest found in San Benito, Texas, for example, contained eight cowbird eggs and two mockingbird eggs. It is assumed that multiple cowbird eggs are the products of several visiting females, not the entire batch of a single bird.

Those who claim that cowbirds are monogamous point out that this behavior suggests they originally cared for their young. Even now, some kinds of cowbirds incubate and brood, but the female loses interest after laying and the male takes over most of the protection of the young. Males in other cowbird species are not protective and have a weakened territorial instinct — and maybe that explains why the female uses foster parents.

-13-

Nestlings and Fledglings

Downy woodpecker

IF ALL GOES ACCORDING TO SCHEDULE, almost before you know it, there'll be babies in the nest. At this point most bird parents have a tremendous amount of work to do. Providing food for their young can require an almost constant stream of activity, plus they must keep the nest clean and be on the lookout for intruders. The necessary amount of brooding — keeping the nestlings warm in the nest — depends on circumstances as well as species.

What their parents must actually provide for the nestlings varies widely. In the case of **precocial** birds — those hatched with down and the ability to walk — the hatchlings find some or all of their own food. For them, the nest is not even a memory.

Altricial birds are born naked and helpless, unable to do much besides gape to receive food brought by their parents. They must wait for the growth of primary wing feathers before they can leave the nest.

LIFE IN THE NEST

For altricial birds, considered less primitive than precocial birds, nest life lasts anywhere from eight days to eight months. The time necessary may be influenced by the temperature and available food as well as the size of the bird and its

Precocial and Altricial

The word **precocial** comes from a Latin root meaning "already knowing." The word **altricial** comes from a Latin root meaning "nursing" or "nourishing."

degree of development at hatching. During good weather and with abundant food, nestlings may leave the nest sooner. Prolonged rainfall, however, endangers newly hatched birds because it may chill them or reduce their food supply.

Parents themselves can present a danger to their young, especially among birds whose hatching times are staggered. The first hatched tend to get the attention, and later "extra" ones may be neglected or even eaten, a practice that occurs among boobies, pelicans, storks, eagles, and owls.

The frequency of feeding nestlings varies with the species. Songbirds may feed their young every half hour or up to forty times in an hour, but a young albatross gets fed only two to five times per week. In general, the smaller the species, the more often it needs feeding, a matter of higher metabolism rates. Passerines may eat the equivalent of half their body weight daily.

Female evening grosbeak with chicks.

A tree swallow feeds a nestling.

With the pressure of keeping those hungry babies fed, territorial male songbirds usually turn their attention from defending the territory to helping feed the young. In some species — notably jays — unmated birds also help out. Interestingly, at first the parents don't seem to recognize their own young (nor do the young recognize them). It's not especially unusual to see a songbird that's carrying food for its own young give it to the first begging baby it sees, even one of a different species.

Most birds keep their nests sanitary by removing fecal sacs (at first the parents may swallow them), but some are notoriously untidy: phoebes and swallows have dirty nests, and doves are particularly poor housekeepers.

PROTECTION AGAINST PREDATORS

In an attempt to keep intruders from their eggs or young, many birds use some kind of **distraction behavior.** One method is feigning injury, as the killdeer does. It drags a wing on the ground, trying to draw predators toward itself as it

Different Feeding Styles

Feeding among the birds is interesting to observe. Some birds feed their young from the **crop,** that saclike feature found in gallinaceous (chickenlike) birds and doves but absent in other families. The crop, where food is stored temporarily before digestion or regurgitation, opens into the gullet, or esophagus. "Pigeon's milk" is the white fatty composition secreted by the walls of the crop.

Rosy finches carry food to their young in cheek pouches opening from the mouth. Hummingbirds put their bills into the gullet of the young and more or less inject the food. Some observers say it looks a bit like surgery in progress. Hawks and owls carry food to their young in their talons.

A killdeer will drag her wing, feigning injury, to lure predators away from her chicks.

moves away from the nest area. Plovers and grouse use similar tactics. Ground doves, too, fake lameness with dragging wings to lure predators away from the nest. When they consider they're far enough from the nest, they fly. Chickadees and titmice have a "snake display," in which, at the nesting hole, they open their beaks and partially open their wings, all the while swaying back and forth slowly. Suddenly, they jump upward, hiss, and swat the sides of the cavity with their wings. It's a startling performance, to say the least.

It's possible that parents urge nestlings to leave the nest as early as possible in the hope of foiling predators, at least to the extent that not all the young can be destroyed at one time.

FLEDGING

Nestlings turn quickly into fledglings. To **fledge** means to grow feathers; when the bird acquires its juvenile plumage or makes its first flight, it is fledged. Feathers can grow a quarter of an inch a day. In a mere twelve days, sparrows have fully developed primary flight feathers.

Fledglings are still quite dependent on their parents and become self-sufficient gradually. The amount of self-feeding or begging that fledglings do depends on how hungry they are. Although you may see fledglings the size of the parent trailing along demanding food, watch closely and you may see that if Mama gives all the tidbits to a sibling, the begging

A female wood duck will teach her ducklings to fish and fly.

bird will hunt for its own dinner for a while. In time the parents seem to get bored with the whole business, and the juveniles are on their own.

RESCUING NESTLINGS AND FLEDGLINGS

If a nestling falls from its nest, it's okay to put it back. The Audubon Society says that the parents won't abandon offspring just because we've touched them. Cover the nest with your hand until everybody settles down. If the nest itself blows down, try to replace it.

With fledglings the situation is different. True fledglings will not stay in the nest if returned to it; they will continue trying to fly. If you encounter a fledgling that seems to be in distress, the best thing you can do is leave it alone. Despite appearances, the parents are probably nearby and will look after it. A fledgling can fly successfully from the ground. You can help by keeping dogs and cats away.

If you move the fledgling somewhere you consider safer, the handling won't prevent its parents from caring for it any more than handling a nestling provokes desertion.

If a bird is truly abandoned (and that's not easy to determine), it needs more than you may be prepared to provide: warmth, protection from direct sun and hard rain, and, especially, frequent feeding.

The feeding gets complicated. Make sure you're willing and able to continue, on schedule, before appointing yourself caretaker of a fledgling. If you choose to make that commitment at some point, remember these tips:

- Feed an omnivorous bird canned dog food or raw beef kidney, bits of hardboiled egg, earthworms, and fruit, using forceps or a small paintbrush
- Give a seed eater earth, charcoal, or crushed seeds to help digest seeds
- Let the bird get its liquid from the food itself. Moisten the food; don't try to put liquid down the bird's throat

What we know about the care and feeding of nestlings and fledglings is patchy at best. There's still plenty of information to be gleaned. If you're interested in helping to gather information, get your binoculars and find a good comfortable site for spying.

-14-

Preparing for Winter

ALL OF THIS FRANTIC BIRD ACTIVITY is geared to the perpetuation of the species; all must be accomplished on a timetable. Not all birds migrate, but all birds in temperate climates must prepare themselves to face the exigencies of the changing season. For the young of species that migrate, it's a matter of adequate maturity for making a long trip; for the young of species that wander or are resident in the home range, it's a matter of adequate maturity to get through the winter. Either way, it comes down to feathers. The young of a given breeding season must have time to grow the necessary body feathers to protect their skin from rain and cold, and they must have flight feathers.

MOLTING

Adult birds molt — that is, they shed one batch of feathers and get a new batch. Some of them molt annually, some more often. In most birds molting is not a phenomenon startling to the beholder; in fact, it usually goes unobserved except in a few fairly obvious cases such as mallards, starlings, and American goldfinches.

Mallards have two molts. In early to midsummer, they lose body feathers (as opposed to flight feathers), which are replaced by **eclipse plumage,** so called because the duck's

Western tanager

113

Starlings undergo a striking change when they molt.

normally distinctive colorings are temporarily absent. During that period, the males are as inconspicuous as the females they closely resemble. (Since the females don't have distinctive colorings in the first place, they aren't candidates for eclipse plumage.) In late summer mallards have a complete molt.

Starlings molt from July to September, and their molt is as startling as that of the mallards. The new feathers, which have white tips, give them a speckled appearance instead of the iridescent sheen of their breeding plumage. Even their bills change color. Males have bright yellow bills, which turn bluish in the breeding season; the light yellow bill of the female is pinkish during breeding. After the molt, the bills are gray.

American goldfinches have a full molt in the fall, when they exchange their bright yellow breeding plumage for the rather somber olive plumage of winter. In the spring they have a partial molt — everything but the wings and tail.

MOLTING PERIODS

BIRD	TIME OF YEAR
Canada goose	Summer
herring gull	March & April*/August & September
rock dove	August & September
chimney swift	Late summer/before migration
northern flicker	July to September
hairy woodpecker	June to August
swallow (most)	Late summer
crow	Summer
blue jay	July & August
black-capped chickadee	July & August
house wren	Late summer
northern mockingbird	Late summer
gray catbird	August
robin	July & August
red-eyed vireo	August & September
common yellowthroat	July & August
house sparrow	Late summer
red-winged blackbird**	July to September
common grackle	August & September
song sparrow	August & September

*partial molt — body feathers only
**retreats to marshes to molt

Many small birds navigate by night, using the stars and prominent landmarks such as mountain ranges to help them navigate.

Most birds lose flight feathers in pairs, one from each side of the body, so their flight isn't affected. In some species, however — ducks, geese, and some other water birds not dependent on flight for their food supply — the molting period is of some danger because the birds lose all their flight feathers. The Canada goose is one such bird, confined to ground or water during its annual summer molt.

MIGRATING

Eighty percent of North America's 645 species of birds move some distance seasonally, including the so-called altitudinal or vertical migration — that is, up and down mountains. We become very aware of birds during their spring and fall migrations because they gather in flocks and are highly visible and audible.

No one knows who first observed the seasonal movements of birds, but certainly the phenomenon has always interested bird watchers. Our earliest surviving writings, including the Bible, talk about bird migration. Aristotle mentioned migration but also insisted that birds didn't migrate, they went into hiding at the end of the summer. Others, he said, transmuted: At the onset of summer, the European robin promptly changed into the European redstart! My favorite explanation of migration is one proposed by an Englishman who said birds flew to the moon (it took sixty days), but since there was nothing to eat there, they went into hibernation.

Allen's hummingbird

Long-Distance Travelers

All of our North American hummingbirds migrate. The ruby-throated hummingbird migrates to and from the Yucatán, at between 30 and 40 mph. The sexes migrate separately — the male arrives first — and move northward as flowers appear. They fly about 25 feet above land or water.

We can chuckle at the ancients' notion of birds hibernating, but some birds exhibit torpid behavior that mimics hibernation. A banded whip-poor-will, for example, was found in the same crevice in the mountains of southeastern California during three successive winters, its normal temperature of 106°F depressed to 64°–67°F. Lesser night-hawks and several species of swifts and humming-birds have been discovered in a similar condition during cold snaps when insects are scarce. The torpid condition conserves energy significantly.

City folk can observe migrations as easily as suburban and rural dwellers. As a matter of fact, the city is apt to be an especially good place to observe migration because small green yards amidst great expanses of concrete and asphalt are havens that can attract large numbers of migrants. Bird watchers in places such as Central Park in New York City may well see more species than do their country counterparts.

Divers — pelicans, gannets, cormorants, herons, storks, ducks, geese, swans — and raptors are all migratory. Many shorebirds are migrants. The black-billed and yellow-billed cuckoos are migrants, and the burrowing owl is migratory in the North. Nightjars migrate, and so do swifts. Chimney swifts fly north in groups of twenty to thirty, but migrate south by the hundreds.

Flickers and yellow-bellied sapsuckers are migrants. Fly-catchers, swallows, and martins migrate, and many thrushes are long-distance migrants. Pipits move south; shrikes retreat irregularly from the northernmost part of their ranges. Most vireos and warblers are long-distance migrants. Orchard orioles leave the North in July, immediately after nesting; north-ern (Baltimore) orioles don't leave until September or October. Tanagers are long-distance migrants.

Why and How

Though it is widely believed that birds migrate to escape cold weather, they're actually almost perfectly adapted to withstand extremes of temperature. They migrate to ensure a food supply. Some birds, primarily insect eaters, start south immediately after the molt following nesting. Others hang around for the fall crop of grains, berries, and fruits.

The migration of some birds is much more conspicuous than that of others. One day it may occur to you that you haven't seen an oriole for at least a week, but it's hard to miss the rowdy congregations of blackbirds that gather in the fall before heading south. The restless movements of birds before migration tend toward the direction of their eventual migration.

Long-distance migration is not necessarily a nonstop marathon. In western Massachusetts, where I live, we begin watching for the Canada geese right after Labor Day. Our location offers what they like — a small lake a couple of miles up the road, hay meadows, and many fields of silage corn. They drift in slowly, and by mid-October our town is playing host to hundreds of geese. Visitors to the lake feed them bread by hand, but when the geese are foraging in the fields they won't permit a near approach. On water

Bird Banding

One way scientists have been able to track the seasonal movements of birds is through *bird banding*. This procedure involves attaching a metal ring stamped with date and location to the leg of a captured bird. If that bird is captured again, the band gives birders information on its movements. This procedure allows the records of millions of birds to be analyzed electronically. Widespread use of banding, by thousands of birders all over the world, amasses mountains of useful information on the movements of birds.

The practice of banding birds to get an idea of their migratory movements started in 1740 in Europe when Johann Leonhard Frisch, a German, tied red strings to the legs of swallows. Anyone else capturing the swallows could find out where they had come from. John James Audubon used the procedure on a couple of phoebes in the early part of the nineteenth century.

they congregate mostly in one large flock, while their forays to the fields may be in large or small flocks. We see them flying often as they perform their various activities. Once the corn is cut they feed well, flying so low over our house to the field next door that we can see their feathers. They leave at the end of November when the lake begins to freeze.

Most birds travel latitudinally — north to south. Some, however, have longitudinal migrations, from west to east; the evening grosbeaks that nest in northern Minnesota and winter in New England are an example. The common ground dove is mostly a permanent resident in the deep South, but it tends to move seasonally toward the coasts.

Mallards

Fly-by-Nights

The migration of some species is inconspicuous not only because it occurs at night but also because night flyers travel at a higher altitude than day flyers. Nocturnal migrants, such as thrushes, wrens, warblers (except the yellow-rumped warblers), and sparrows, tend to be relatively weak flyers; by day, they feed on the ground or in plants, as usual. Strong flyers that feed on the wing, like swallows, are diurnal, or daytime, travelers. One day swarms of them infest the telephone wires; the next day they're gone. Hawks, strong flyers that normally have a large range, also migrate during the day, as do geese and other large birds. Geese also travel at night. Ducks and shorebirds are nocturnal, or show no particular preference.

Dangers En Route

Travel Speed

Radar has helped determine the speeds at which birds travel. Most songbirds travel at about 30 mph, ducks can travel at almost 60 mph, sandpipers have been clocked at 110 mph during migration, and swifts hold the record at up to 200 mph!

Once on their way, the travelers frequently experience difficulties. Though running into human-built structures causes thousands and thousands of deaths, weather-related deaths are far more devastating. Sometimes birds are observed in the grips of "reversed" migration — flying in the wrong direction. Small birds are known to travel at about the speed of the wind, and to migrate downwind. Consequently, if the wind is going in the "wrong" direction, so are the birds. Many birds perish in early snowstorms. Birds traveling along a coast may be blown out to sea. Costly as it is, the semiannual trek proceeds on schedule, a phenomenon fraught with mystery.

Navigating

Perhaps the most remarkable aspect of migration is simply that birds somehow manage to reach their destinations,

California's mountain quail migrates altitudinally, rather than north to south. It nests at about 10,000 feet in the Sierra Nevada and then walks down to about 5,000 feet to spend the winter.

often after traveling thousands of miles. People, armed with maps and having the additional advantage of road signs, get lost in their own communities all over the country. Theories abound, but no one has been able to figure out how birds navigate. How do they know where to go? In this century, many carefully controlled studies have investigated the question.

One theory on bird navigation is that birds orient themselves visually by celestial navigation, using the positions of the sun, moon, and stars for guidance. Other experts believe that birds use landmarks. If so, something else must also be at work that helps birds navigate in overcast weather. Radar tracking has shown that birds have reasonably good orientation under, inside, and between cloud layers.

Plenty of other theories have been offered. Many biologists believe that birds use the earth's magnetic field and the mechanical effects of its rotation for orientation in navigation. In support of their idea, they note that birds don't migrate across the Poles.

Others think that orientation is guided by thermal radiation. Since there is less in the north and more in the south, birds seek increasing or decreasing amounts as they fly, depending on the direction in which they're moving. Homing is another possibility.

There are many theories of migration, supported by studies of such complexity that the ordinary garden-variety bird watcher is left floundering in too much information. Among the newer ideas are genetically influenced orientation, encoding of the route itself at hatching, possession of an "internal clock," and learning from experience (young birds are more apt to go astray). There is also a theory that migration patterns are being altered as our climate slowly changes.

Flyways

In North America, birds follow four major routes in returning to their breeding grounds. Some journey from South America north through California and Texas, distributing themselves to the North and West. Another group migrates primarily along the Pacific Coast. On the other side of the continent, birds travel from the tropics through the West Indies and Florida, north along the coast and west to the central states or north to New England and Canada, often along the Hudson River and over Lake Champlain. The fourth and most important route runs from the Yucatán peninsula across the Gulf of Mexico to the Mississippi Valley and points north.

Flying South

The fall migration from north to south lasts more than twice as long as the spring one in the other direction. Along the shore it may begin in July. It will end sometime in November with the departure of finches and sparrows.

A feeding station doesn't usually tempt southbound migratory birds to refrain from their normal travel, although it may permit the survival of those members of a migratory species that for some reason (injury perhaps) were unable to leave with their cohorts. And there is evidence that the proliferation of feeding stations has extended the range boundaries of some species.

Magnetism

In addition to the stars and major landmarks such as mountain ranges, rivers, and lakes, many birds use the earth's magnetic field to help them navigate.

Good Weather for Migrating

Most small land birds migrate by night, and they prefer clear nights with northerly winds. Foggy weather or heavy overcast and unfavorable breezes can lead them far off course. They may sometimes be seen going in the wrong direction until the weather clears and they can reorient themselves. Birds unusual to a given area occasionally come to feeders there because of navigational errors.

Normally, because natural foods are abundant, fall may be a slack time at many feeding stations despite the hordes of migrants passing through an area. If birds elect to stop in your yard, it's entirely possible that water will be the principal attraction. Nevertheless, keeping the feeder stocked, even for a small coterie of birds, is a good idea. The presence of this faithful nucleus is likely to attract other birds later on.

Flying North

In contrast to the southern migration, those birds making their way north in spring are quite apt to visit feeding stations because early in the season natural foods are at their lowest point. Normally, the spring migration lasts only about two months. Day length seems to trigger migration. Perhaps not surprisingly, the first birds to return are those that traveled the least distance.

Young birds making their first trip north don't necessarily return to the exact place where they were hatched, but once birds have nested, they're likely to return to the same area again and again. The young of colonial birds *do* tend to return to the area of the colony in which they originated, however.

Normally, birds of different ages migrate separately. Later arrivals may not be able to find a suitable habitat in their usual summer range and so they move on. One advantage of this is that whatever the existing food supply is, the birds are spread over a wider territory at any given time, so the young get the energy they need to mature faster.

WANDERING

Not all large-scale bird movement constitutes migration. We may think of migration, simply defined, as a regular seasonal movement of birds between widely separated areas. We can distinguish **sedentary birds,** which are permanent residents in their habitats; **nomads,** which wander following favorite food supplies; **middle-distance migrants,** which move only as far as they must to ensure a food supply; and **long-distance migrants.**

Sedentary birds include game birds, pigeons and doves, some owls, and most woodpeckers. Crows and jays are mainly sedentary, but they generally retreat from the northernmost parts of their range in winter. Starlings are sedentary in North America (European starlings migrate), and so are weaver finches, or house sparrows. Some blackbirds are sedentary;

Pileated woodpecker

some drift southward. Buntings, sparrows, and cardinals may be sedentary or migrants.

Some species, such as cedar waxwings, are considered permanent residents of an area even though after nesting they may wander considerably, following a favorite food supply; this is not considered migration. Great horned owls, grouse, and golden eagles can be found most of the time throughout their permanent ranges, though they sometimes shift their territories to find adequate food supplies — a sharp-tailed grouse unable to satisfy its appetite in North Dakota may wander west to Montana, for example. A species of Asiatic grouse once moved into western Europe and Great Britain in search of food, later returning to its former range. Nomadic birds in Australia move around strictly according to the food supply. In winter it may be only a matter of moving to a lower altitude in the same area, known as **altitudinal migration;** every thousand feet lower is the equivalent of going three hundred miles south.

Some birds travel great distances during the course of a season to follow a favored food supply. The failure of a crop may send them packing. But because birds of different species have varying food preferences, the failure of any *one* given crop isn't likely to cause the disappearance of all birds from that area.

The goshawk and snowy owl, resident in the Arctic, move south whenever their food supply becomes poor. The lemmings that form the bulk of the snowy owl's diet suffer sharp declines in population about every four years, so the owls move southward to prey on other small mammals. Even among those birds thought of as migrators, some individuals may remain in or near their breeding areas.

Some birds — the cardinal is a prime example, but goldfinches and purple finches are known to engage in the activity, too — just seem to wander on a whim for no discernible reason. We've become aware of these vagaries thanks to the procedure of bird banding, which enables ornithologists to keep track of the whereabouts of individual birds within a species.

Gulls, most owls, waxwings, and some finches are nomadic birds. The kingfisher, resident in the north during the breeding season, migrates only far enough south to find the open water it needs. Titmice, chickadees, nuthatches, and creepers drift southward, too. If you have them year round, your summer residents may be a different batch from your winter residents. Wrens seldom migrate far. Mockingbirds, catbirds, and thrashers usually retreat to the south. Robins are middle-distance migrants.

Mourning dove

Making History

It was the captain of the *Pinta* who noticed a group of land birds far out at sea and persuaded Columbus to shift his course to the southwest. Had he maintained his original course, Columbus would have landed on the mainland instead of on San Salvador.

Nomads

The young of many species wander, sometimes extending the range of the species. For example, most jays that actually migrate are the young; the adults usually stay together in a flock as a residential community.

Migration after the breeding season stretches over a long period of time. For birders, it's a time of excitement, a time to see species not often easily observed. It's a poignant time, too. Though some birds disappear virtually unnoticed, the flocking that occurs in late summer and early fall is a sharp reminder to gardeners and lovers of the sun that winter is coming. Even if you enjoy the change of seasons and look forward to the coming of a new one, some atavistic tremor of apprehension accompanies the departure of the migrants. A wedge of geese is exciting and exhilarating whenever one sees it, but in the fall it seems somehow less joyous. Maybe it's partly because it signifies the end of beginnings, a farewell. On the whole, it's simply more pleasant to welcome the travelers back than to see them go.

The northern saw-whet owl, the tiniest of North American owls, is a useful hunter of rodents.

PART III

INDIVIDUAL BIRDS

The following section contains closeup looks at more than 135 individual species of birds, organized by family. Not all of these are feeder birds, by any means; we have included a number of birds such as woodpeckers and raptors that you might see feeding or nesting in your vicinity.

In the reference chart that accompanies each species, "L" means the length of the bird, measured from tip of bill to tip of tail, and "W" means wingspan. The breeding and winter ranges are approximate, and an individual bird might show up far from its customary range.

For more information on preferred feeder food, habitat, nesting habits, descriptions of eggs, and nesting box dimensions, see the charts in the Appendix.

-15-

Getting to Know the Birds

An INTEREST IN BIRDS is insidious, creeping up and catching you unawares. What I remember about birds from growing up in western Pennsylvania is cardinals and robins, pheasants and sparrows. Mother was forever telling me to come to the window to see the tufted titmouse feeding or bidding me look in the spiraea to discover the catbird. I was aware that she thought highly of both, but if the bird wasn't brightly colored, or ubiquitous like the house sparrows, I didn't remember it from one minute to the next. Later, in all the places my husband, Bud, and I lived, there were probably hordes of birds, maybe even lots of different species. I never noticed.

Then we moved to western Massachusetts and began to garden on a half-acre lot in a semirural area. There seemed to be an inordinate amount of bird activity. Grackles screamed from the white pines. Chickadees flittered through the dogwood branches. Downy woodpeckers drummed on the elms. A flock of evening grosbeaks descended on the apple tree one winter morning, looking like exotic tropical birds against the background of snow and dazzling sky and bare branches. It was a veritable revelation; I was hooked. I went out and bought a copy of Peterson's *Field Guide*. Enough of this "yellow birds with black wings and white patches and beaks like parrots."

Evening grosbeak

Maybe you've reached that same stage. You've provided food and water, and the birds have discovered your efforts. You don't have any trouble recognizing the robin flitting around the birdbath. Or the blue jay busily — or greedily, depending on your point of view — carrying away sunflower seeds. But what about that little brownish bird? Sparrow, you think? But aren't there roughly ninety-three different *kinds* of sparrows? Or maybe it isn't a sparrow at all. Maybe it's one of those *other* little brownish birds.

Get Help

That's the state of the art with many of us. So what's the first step? A friend who is an experienced birder is surely the best of all possible aids in learning to identify birds. But even if experienced guidance is available to you, there will certainly be times when you must have recourse

---

Observing Birds

Make mental notes of all the physical characteristics you notice while closely observing a bird, and *then* look for a matching picture in your guide. Ask yourself these questions:

- Does it show any splotches of color?
- What kind of beak does it have?
- What kind of a tail?
- Does it have a crest or crown patch?
- Any discernible pattern such as stripes or rings?
- Any markings around the eye or head or back or breast?
- In flight do you notice any particular characteristic, like gliding or dipping or undulation?
- Do any flashes of white or some other color appear on wings or tail when it moves?
- Where did you see it?
- What about its shape — is it chunky or slender?
- Does it walk, hop, perch quietly, or flit about?
- Does it say anything?
- What about size? Give yourself some mental image to compare the bird to. Was it about the size of a sparrow (5 to 7½ inches)? A robin (8½ to 10½ inches)? A crow (17 to 21 inches)?

---

to other methods. You'll want to know how to go about the process all by yourself.

In our household, books accumulate as readily as dust. What we consider the indispensable volumes for identifying birds are Roger Tory Peterson's *A Field Guide to the Birds* and *A Field Guide to Western Birds* (Boston: Houghton Mifflin Company).

The first thing to do is to familiarize yourself with the books. Browse awhile. You'll notice that the individual pictures of birds have black lines pointing to those conspicuous characteristics differentiating one bird from another.

Now, obviously, you don't intend to memorize every plate in the book. And if you go rushing for your field guide when you see an unfamiliar bird, chances are it'll be long gone by the time you find either the guide or the page in it that you want. So look at the bird, while you have the chance.

You may find the written descriptions of song helpful. They mean nothing to me until after the fact of having heard and identified the song. In either case, a recording of bird songs may prove invaluable. Your local library may have one, or be able to get you one. A number of excellent recordings are now available.

Magpie

BINOCULARS NEEDED

What you need, of course, are field glasses or, better still, binoculars. Actually, binoculars are easier to find, unless you happen to stumble across field glasses at a tag sale or flea market. Binoculars differ from field glasses (or opera glasses or a telescope) in that prisms have been added to the magnifying lens or series of lenses. You'll be able to use them to advantage both earlier and later in the day, and what you're hunting will be easier to find because their field of vision is wider than that of a magnifying lens.

What does that mean? Focus your binoculars on the top of a telephone pole, say, a hundred yards away. Should you try the same thing with field glasses or a telescope, you'll notice that you can see farther to each side of the pole with the binoculars than you can with field glasses or the scope. That's important when what you're trying to locate has a tendency to move around a lot, and is not apt to oblige you by perching right next to something you can sight in on easily.

Binoculars come in several degrees of magnification, the usual being six, seven, or eight. You can get ten or twenty power, but they're more expensive and somewhat harder to use.

You'll find a bewildering variety of binoculars available, in a wide price range. Birding and nature magazines often run articles comparing different binoculars.

One word of caution about binoculars. Sometimes you'll find advertised what sounds like a great price. The ad may say "TEN TO FIFTY POWER MAGNIFICATION!" Read the fine print. If it says they're "non-prismatic," they're not really binoculars, no matter what they look like. Non-prismatic binoculars are a contradiction in terms. Binoculars aren't straight tubes. Those bumps contain the prisms. You'd be better off with good, optical quality field glasses or a telescope than with "non-prismatic" binoculars with molded or poured plastic lenses, and phony bumps.

Although binoculars are easiest to use, field glasses and telescopes you might happen to have around the house can also be handy gadgets. The other day we were trying to identify an uncooperative bird on a branch, and our 8×40 binoculars simply didn't pick up the necessary detail at the range. Bud's spotting scope, at thirty power, did the trick. Especially in your own garden on a bright day, you can become quite adept at picking out a particular branch simply because you're so familiar with the setting. One of the problems with the scope, however, is that the higher the magnification, the harder it is to hold it steady on the object observed.

That's where a tripod comes in handy. If you have a place to keep a scope mounted near a window, it'll be even more useful. Honesty urges me to add that whichever window it's close to won't be the window from which you spot the bird. The same principle holds true for binoculars or field glasses. Ours hang on a nail beside the west kitchen window. Birds unidentifiable with the naked eye are invariably sighted from the east kitchen window, unless someone has carelessly left the binoculars on the table beside it. It's a rule.

One other possibility is a monocular, which is essentially half of a binocular. Still prismatic, it's used by one eye only. Its drawback is that you can't achieve the three-dimensional quality you get with binoculars.

Next Steps

Once equipped with field guide and binoculars, you might want to go one step further and take a course through adult education at a nearby school. If your interest continues to grow, you may be tempted to investigate local bird clubs. One way or another, you're all set to identify your guests. A lot of our fun in feeding birds consists of watching their be-

American kestrel

havior, and the field guides give few clues about that. If we're to have the chance to observe characteristic behavior, we'll also need to know just what foods to use to entice a particular species to the feeder. Discovering that preferences are for berries or insects or seeds just isn't enough. Which berries? What do I substitute for insects — or won't those birds come to the feeder at all? What kind of seed?

FORM AND FUNCTION

Once you acquire the habit of looking closely at birds so that you can identify them by referring to your field guide, you'll discover that some of their physical characteristics provide clues to their behavior. Some of these physical attributes are related to the feeding behavior of birds. Take the size and shape of the bill. Seed eaters like sparrows and finches have short, stubby ones. Cardinals and grosbeaks, with their thicker, heavier bills, can handle larger seeds than, say, a goldfinch. The scissorlike bills of crossbills enable them to extract seeds from pine cones. Woodpeckers have strong bills, like chisels. Warblers and wrens, with their thin, sharply pointed bills, are primarily insect eaters. Birds of prey have hooked bills, ideally suited for tearing meat.

We've mentioned that birds seem to have color preferences and, in passing, pointed out that they have superb vision, perhaps the sharpest of any animal. Obviously that helps them locate food and contributes to their safety. The position of the eyes on the head is a hint to whether the bird is the hunter — hawks and owls look ahead — or the hunted — woodcocks and grouse have their eyes positioned more to the side and can see farther to the rear than can birds of prey.

Hearing is acute in birds too. Woodpeckers are thought to listen for insects, and owls rely on sound when hunting their prey. Experimental studies have determined the range of hearing in many species. Though many of us may have believed that robins locate earthworms by sound, ornithologists

The "Edge Effect"

Although you can travel widely to observe birds, you don't actually need to leave your own neighborhood. Your immediate area will provide a plethora of birds during the different seasons, according to the habitat each species requires. Look for areas of transition from one kind of plant community to another; they tend to attract more birds of different species than do either of the separate communities. On the fringes of woods, for example, forest-dwelling warblers mingle with chickadees commonly seen at our feeders and in our shrubs. At the edges of swamps, shy woodcocks compete for earthworms with robins content to nest over our front doors.

Bill shape is a clue to birds' feeding behavior.

tend to think that the characteristic cocking of the head is related to visual hunting instead.

Look at the legs and feet of the birds you see for further clues to their habits: webbed feet for swimmers, long legs for waders, talons for birds of prey. Birds such as swallows and hummingbirds that take their meals on the wing have small, weak feet. They can perch, but their feet aren't much good for anything else. What a contrast to the thrushes, which forage on the ground!

Color is another hint about behavior. Ground dwellers tend to be duller in appearance than arboreal (tree-inhabiting) birds. Sparrows, which usually prefer to nest and forage on the ground, are brownish. Grosbeaks and finches, mostly brightly colored, are more active in trees. (There is evidence to suggest that the brightly colored birds aren't as tasty to hunters, either.) Females and juveniles generally tend to be less colorful than males even among arboreal species, doubtless another natural safety device to provide protection for incubating and other relatively helpless creatures.

A PLENITUDE OF BIRDS

There are over 8,500 *living* species of birds in the world, and it's estimated that the present bird population is about 100 billion. This continent itself has nearly 950 species in seventeen orders and more than seventy families. We can't cover them all. Choices must be made. No matter how that's done, somebody's favorite will get short shrift or be ignored completely. What I'll attempt to do is nod in the direction of the more common feeder birds to get you started.

If the brief capsule accounts whet your appetite for complete biographies, help is at hand. In 1960, the highly respected John K. Terres started work on an encyclopedia of this continent's bird life, the heart of which is biographies and color pictures of 847 species that breed or have been sighted in North America. The scope of the work is impressive indeed. The 1,109 pages of *The Audubon Society Encyclopedia of North American Birds* were finally published by Alfred A. Knopf in 1980. Twenty years were devoted to the project, and it represents our definitive study of birds.

It is by no means the only work available, however. Another outstanding work is the *Birders' Handbook: A Field Guide to the Natural History of North American Birds* by Paul Ehrlich et al. (Simon and Schuster, 1988). You'll be able to find impressive and/or entertaining bird lore whatever your area of greatest interest or your taste in reading material.

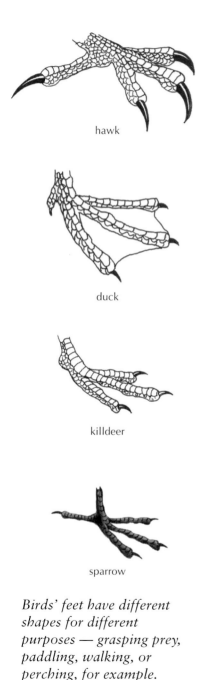

hawk

duck

killdeer

sparrow

Birds' feet have different shapes for different purposes — grasping prey, paddling, walking, or perching, for example.

LIST OF BIRD SPECIES

(list continued overleaf)

WATER BIRDS

MANY of the water birds are not hard to attract — a mixed blessing, maybe. Almost any city park with a body of water in it has geese and ducks, wild and domestic, that are as hungry for handouts as any pigeon. **Mallards** (ancestors of most domestic ducks) and **Canada geese** are found all over the country. They are all members of the order Anseriformes, family Anatidae. Ranging from medium-size to large, all of them have webbed feet and bills that serve as strainers.

Sometimes, however, when these geese and ducks find a likely looking place, they forget all about migration. In their enthusiasm to provide suitable habitat for geese, some cities and towns — some states, even — have managed to create king-size problems for themselves. You think pigeon droppings are a nuisance? Geese are much worse. They become half-tame very

willingly if ample food is available. Some people, and some government agencies, plant grain for the visitors. But geese can't — and don't — discriminate between what's planted for them and what's planted for market. They're big birds with enormous appetites. A flock of them can strip a field in short order. Gleaning is one thing, and many farmers welcome migrants stopping for a snack *after* the harvest. But what has happened in some areas is that geese, both imported and migrant, have found the living so fine, they've settled in permanently. They raise families, which in turn . . . but you see how it goes. Canada geese are now breeding much farther south than ever before, and farmers' grain fields have begun to suffer.

The same thing can happen on a small scale — to you. Having raised domestic geese and ducks, I'm painfully

aware of the extent of their appetites. We don't feed the wild ones. We're glad a pair of mallards nest at the brook behind the back pasture. We're delighted to see the Canada geese stopping in the cornfield next to us on their way south. I know of no avian event more exciting than the migration of geese. At the first honk, I drop whatever I'm doing and rush outdoors to watch until they're out of sight and hearing. What splendid creatures they are! However dreary the day, seeing them exhilarates me. But they're going to have to hustle for themselves. Enough is enough.

Swans, loons, and geese mate for life, and for birds they live long lives. The **mute swan** can reach the age of 100 years.

Ducks tend to have more than one partner. As with pheasants, the females have the job of rearing the young. The drakes are frequently almost gaudy in their nuptial plumage.

Left to right: Mute swan, herring gull, Canada goose, wood duck, mallards, common loon.

Common Loon

(Gavia immer)

WITH its bold black-and-white plumage, cautious nature, and haunting call, the common loon has become a symbol of the northern wilderness. Loons live on remote freshwater lakes in the forests of Canada, Alaska, and the northernmost tier of the United States. If you are lucky enough to have a pair nesting nearby, you probably already know not to disturb the two large, olive-brown, black-spotted eggs — which may be sitting exposed on a rocky beach. An intrusion, especially by a power boat or water skiers, might cause the loon parents to abandon the eggs.

Loons keep the same mates throughout their long life of twenty-five years. They arrive at their breeding grounds in spring as soon as lakes thaw, each pair claiming a watery territory of 20 to 100 acres.

They need that much room both for fishing and as an airstrip. The loon's method of becoming airborne is to run in a noisy flurry along the lake surface until it can gain sufficient momentum to rise on its stubby wings. When startled, it would much sooner dive than take to the air.

Loon nests are most likely to be found on small wooded islands. They consist of a loose collection of rushes, grass, twigs, and reeds, and may be on a floating bog or a muskrat house — or the parents may simply scrape a shallow dish right at the water's edge so they can slide directly from nest into water at the slightest sign of danger. Loons are practically helpless on land because their legs are positioned so far back.

The fluffy brown chicks can swim as soon as their natal down dries. It takes several months of teaching by both parents, however, for them to learn the finer points of fishing and flying. When the chick gets tired one of the parents will hoist it onto its own back.

Loons often appear first in bird books because they are, evolutionarily speaking, so primitive. Diving birds in general are the most closely related of all birds to reptiles, from which they are supposed to have descended, and loons in particular are several million years old.

Size: 32"L
Breeding range: Alaska and northern Canada south to central Massachusetts, west to Montana and Washington
Winter range: Atlantic coast from Newfoundland to the Gulf of Mexico; also along Pacific coast south to Mexico

Preferred habitat: Freshwater lakes
Preferred nest site: On bare ground in shoreline vegetation, generally on a small wooded island, floating bog, or muskrat lodge
Clutch size: 1 to 2

Incubation period: 28 to 29 days
Nestling Period: 1 day (precocial)
Broods per season: 1
Preferred food: Fish, amphibians, insects, aquatic plants

Mute Swan

THE mute swan, an introduced species established along the Middle Atlantic coast as a resident, is found frequently as a more or less domestic ornament to ponds and rivers. Measuring sixty inches from bill tip to tail tip, it's the swan that most of us are likely to see.

Swans belong to the same family as ducks and geese, but you're not likely to observe native swans at close range. Just seeing them passing aloft is a rare treat for most of us. Our native swans are the trumpeter (which at 30 pounds and measuring up to 72 inches is the largest swan) and the tundra (formerly called whistling) swans. Trumpeter swans, in the Northwest,

are increasing in numbers, but the tundra swan is our most common native swan. It winters usually from Chesapeake Bay to North Carolina and from southern Alaska to southern California, with some found in the Great Basin, the lower Colorado River, and New Mexico. In other areas it is chiefly a migrant.

Swans have notoriously short fuses, especially during the nesting season, so be careful. A blow from those powerful wings is no joke, and they're proficient in the use of their beaks as weapons. Nesting swans sometimes attack and kill other birds (ducks, for instance) that inadvertently venture too near their nests.

Size: 60"L
Breeding range: Wild populations established along northeastern coast from New Hampshire south to Virginia; also northern Michigan, northeast Wisconsin, northeast Illinois, northwest Indiana, and south-central Montana

Winter range: Resident in breeding range
Preferred habitat: Protected coastal waters with dense aquatic vegetation
Preferred nest site: On ground, secluded; sometimes colonial
Clutch size: 2 to 11, usually 5 to 7

Incubation period: 35 to 36 days
Nestling period: 1 day (precocial)
Broods per season: 1
Preferred food: Crustaceans, insects

Canada Goose   *(Branta canadensis)*

CANADA GEESE defend an area of a quarter acre to an acre until the time of hatching. Like a swan, a Canada goose defending his territory is not to be taken lightly. As a mild threat, he lowers his head at his opponent and hisses. If intending to attack, the male calls *ahonk* (the female's version is simply *honk)* and rapidly raises and lowers his head. Very aggressive behavior preceding an attack consists of waving his head and neck back and forth; interestingly enough, such behavior is also part of the greeting behavior of mates.

Canada geese pair for life before sexual maturity and, in addition, behave as if they grieve the loss of a partner. In courtship they stand side by side, honk, and curve their necks. It doesn't sound like much, but it satisfies the geese.

The nest is merely a depression lined with sticks, grass, reeds, and down, which the female builds on the ground near water. The hatchlings are precocial, so they're out of the nest shortly after hatching. However, down is not the protection that feathers are, so the parent bird broods the goslings at night and for a few hours during the day. As fledglings, the young depend on the parents only for protection.

Size: 16–25–45"L, 50–68"W
Breeding range: Arctic coast south to South Dakota and the Gulf of St. Lawrence, south along the Atlantic coast to North Carolina; Canada except South Ontario and Quebec and U.S. from Arctic coast south to southern U.S.

Winter range: Southern Great Lakes and Nova Scotia to New England, south to the Gulf of Mexico and Florida, southeastern Alaska to California
Preferred habitat: Ponds, rivers, bays, fields, saltwater marshes
Preferred nest site: On the ground, near water

Clutch size: 4 to 10, usually 5 or 6
Incubation period: 28 days
Nestling period: 1 day (precocial)
Broods per season: 1
Preferred food: Grasses, marsh plants, aquatic plants, grains

Mallard

(Anas platyrhynchos)

MALLARDS engage in group displays before forming pairs, courtship beginning in the fall and continuing until spring. Males shake both heads and tails, arch their necks with heads pointing to the water, and whistle. Both males and females engage in mock preening. Immediately preceding copulation, pairs face each other, bobbing their heads. Forced copulation is common among mallards during both incubation and fledgling care.

A pair of mallards typically needs an eighth to a quarter of an acre of feeding territory. The male defends only a small area of water and reeds, not including the nest, for ten to fourteen days until incubation begins.

Mallard nests are depressions in dry ground, usually near water and either in woods or brush. Sometimes they are found in tall grass or in alfalfa fields. The inside diameter of the depression is 8 inches, and the nest is made of grass, leaves, and reeds. The female selects the site and builds alone, starting to lay

her eggs on bare ground and gradually assembling more and more nesting materials, constructing her nest as she incubates and placing down over the eggs when she leaves them.

Mallards stay in the nest a day or less; on the second day the ducklings are led to water. The fledgling period lasts fifty to sixty days.

Size: 23"L, 36"W
Breeding range: Western North America east to the Great Lakes and New England, south to northern Virginia, and west to New Mexico, Arizona, and California
Winter range: Western coast of North America east to Massachusetts, south along the coast; inland from southern

New England west to southeastern Alaska, south to southern Mexico; permanent resident in the middle part of its range
Preferred habitat: Edges of lakes, ponds, reservoirs; sometimes grasslands and fields far from water
Preferred nest site: In tall grass, dry ground

Clutch size: 6 to 15, usually 8 to 15
Incubation period: 23 to 30 days, usually 26
Nestling period: 1 day or less (precocial)
Broods per season: 1
Preferred food: Seeds, leaves

Wood Duck

(Aix sponsa)

THE drake of this species is the fanciest of North American ducks, sporting iridescent feathers during the mating season. These beautiful birds, which accept nesting boxes, nest in a tree cavity from 3 to 60 feet up, in wooded swamps or marshes or beside water. The cavity is lined with wood chips and down.

Wood ducks have sharp claws that enable them to climb up from the nest to the entrance hole, a distance of 4 to 8 feet. Some observers believe the female carries the young to water in her bill, but most think the parents coax or urge them to leave the nest and then lead them to water.

Size: 18½"L, 28"W
Breeding range: Western coast of North America from southern British Columbia to central California, northern Nova Scotia south to south Florida, west to the midwestern states
Winter range: Maryland, south to Florida and west to central Texas; coast of Washington south to central Mexico
Preferred habitat: Wooded swamps, marshes, near bodies of water
Preferred nest site: In tree cavity, nestbox
Clutch size: 10 to 15, usually 6 to 8 (another source says 6 to 15, usually 9 to 14)
Incubation period: 28 to 31 days (another source says 28 to 37, average 30)
Nestling period: 1 day (precocial)
Broods per season: 1
Preferred food: Acorns, insects

Killdeer   *(Charadrius vociferus)*

ALTHOUGH not a feeder bird, the killdeer often nests — if you can call it that — near humans. This long-legged plover of the family Charadriidae seems to have an utter disdain for nest building. While other birds gather quantities of materials and construct elaborate homes, the killdeer lays her three or four eggs right on bare ground, a flat gravel rooftop, or even a gravel parking lot. She then vigorously guards them against passersby. Fortunately, the chicks are precocial and can get out of harm's way — usually to a nearby marsh or brook — immediately upon hatching.

If an intruder threatens to disturb the nest, the killdeer will run out, dragging its wing and feigning injury to provide a distraction. Other birds and animals use the killdeer's alarm note as a warning. It is an excellent flier whose speed can reach 55 mph.

Size: 10½"L
Breeding range: Southern Canada and southeast Alaska south to Mexico, coast to coast
Winter range: Southern New England west to California and north along the Pacific coast to southwest British Columbia, south to South America

Preferred habitat: Fields, parks, open areas, often close to human habitation and water
Preferred nest site: Bare ground, also graveled rooftops
Clutch size: 3 to 5, usually 4
Incubation period: 24 to 29 days

Nestling period: Less than 1 day (precocial)
Broods per season: 1, sometimes 2
Preferred food: Insects

Herring Gull

GULLS (order Charadri-iformes, family Laridae) think bread and potato chips are wonderful, but these foods are not recommended. If you want to try something more nutritious, gulls get a little fussy. Meat or fish scraps, fine, but forget birdseed. People whose houses have tiled roofs sometimes curse gulls who have learned that a roof is a dandy place for dropping shells to open them. Unfortunately, the shells tend also to open tile roofs.

The colonial herring gull defends an area of only 30 to 50 yards in diameter, chasing or fighting intruders at its boundaries. Newly mated pairs patrol the periphery. While courting, herring gulls engage in mate feeding and head tossing. After finding a

mate, the male regurgitates food for the female. Perhaps this ritual feeding of the female of various species developed from the female's increased caloric needs at breeding time.

The herring gull nests in open, sandy areas on the ground. Both parents scrape a shallow depression and then build a nest of grasses, sea-

weed, shells, and feathers, with an inside diameter of 8 to 10 inches.

Herring gulls hatch over a period of three days or more. The female broods for the first few days. Both parents feed the young, by regurgitation, for five weeks. The most common predators of both nestlings and fledglings are other gulls.

Size: 25"L, 55"W
Breeding range: Arctic; northern Alaska except North Slope, Mackenzie area; south along Atlantic to South Carolina; Pacific coast to central British Columbia; and larger inland lakes to Minnesota

Winter range: Newfoundland south and west to Great Lakes, south to Gulf of Mexico; southern Alaska along Pacific coast to Mexico
Preferred habitat: Shores, farmland, dumps
Preferred nest site: On ground in open sandy area

Clutch size: 2 to 3
Incubation period: 26 days
Nestling period: A few days
Broods per season: 1
Preferred food: Carrion, garbage, marine animals

THE ORDER OF FALCONS

PREDATORS are traditionally a contingent we take care to bar from the feeder. But raptors, many of which are now endangered, are important and beautiful members of the ecosystem, so it's useful to know more about their habits.

Birds of prey include both day- and night-flying raptors. That word "raptor" means *any* bird of prey, and I use it to avoid offending birders who quite rightly wince when they hear all day-flying raptors lumped together as hawks. Not all birds of prey belong to the same order, let alone the same family. Owls belong to the order Strigiformes. Falcons, accipiters, buteos, eagles, and vultures are found in the order Falconiformes. And the order of perching birds, Passeriformes, contains a predatory family, the shrikes.

If you maintain a feeding station, you'll occasionally see songbirds taken by birds of prey. No need to get agitated. There was a time when misinformed bird lovers felt obliged to wage war on birds of prey. No question about it, it's a shock to the average bird-lover's nervous system to see a Cooper's hawk strike a robin on the lawn. We have to remember that all these birds that sometimes prey on songbirds also prey on rodents. They're a positive boon to agriculture. If that's still not good enough to help you to accept the situation, remind yourself that it's no "worse" than to see that same charming robin dining on a grasshopper. We tend to become upset about the whole business whenever the animal food of some creature we admire happens to be one of the higher animals, but it's not a reasonable attitude. Bear in mind that there are relatively few vegetarians among human beings.

I'm not suggesting that you'll want to *provide* dinner for birds of prey. Occasionally an injured one will visit your feeder. Occasionally you may see a hunter surveying the situation. The former is no menace, and the latter isn't likely to capture any but diseased or injured birds. You may discover that the heightened appreciation of songbirds that accompanies having a feeder in your yard will gradually grow into an interest in other kinds of birds.

Left to right: Bald eagle, red-tailed hawk, black vulture, turkey vulture, northern harrier, American kestrel, osprey.

Turkey Vultures
Black Vultures

(Cathartes aura)

(Coragyps atratus)

AS carrion eaters, these vultures perform an important service. Turkey vultures are larger than black vultures and have naked red heads; black vultures have black heads.

With nearly the wingspan of eagles — six feet — turkey vultures are splendid flyers that often soar in groups of thirty or more. They have dramatically extended their range northward until they now breed as far north as southern Canada. Their diet consists mostly of carrion; their acute sense of smell helps them find dead animals from high in the sky, where they ride warm air currents on widespread "finger-tipped" wings. They roost in evergreen groves and lay their eggs on the ground, in gravel, or in sawdust.

The black vulture is smaller than the turkey vulture, with a wingspan of about five feet. It flies differently, too — less effortlessly, with short hard strokes alternating with brief glides. Black vultures don't bother with a nest at all; eggs are laid in hollow stumps, thickets, and caves.

Turkey Vulture

Size: 28"L
Breeding range: Southern New England west to the Great Lakes, thence through southern Canada to British Columbia, south to South America
Winter range: New Jersey and Maryland west through Ohio to southern California and south
Preferred habitat: Various, including wooded and open, wet and dry
Preferred nest site: On the ground, among rocks or gravel; in sawdust in hollow tree trunks
Clutch size: 1 to 3, usually 2
Incubation period: 30 to 41 days
Nestling period: 70 to 84 days
Broods per season: 1
Preferred food: Carrion

Black Vulture

Size: 25"L
Breeding range: Mason-Dixon line to Missouri and south through Texas and southern New Mexico and Arizona and south
Winter range: Permanent resident in most of breeding range
Preferred habitat: Various
Preferred nest site: Hollow stumps, thickets, caves
Clutch size: 2
Incubation period: 28 to 29 days
Nestling period: 67 to 74 days
Broods per season: 1
Preferred food: Carrion

Bald Eagle

(Haliaeetus leucocephalus)

DURING courtship the male bald eagle performs aerial loops for the female, an unforgettable sight. The pair builds its aerie together, in the fork of a giant tree or on a high cliff ledge. A huge pile of sticks and branches lined with grass, moss, sod, and weeds, the nest starts out being 2 feet high and about 5 feet in diameter. The pair may use the same nest for years and continue adding to it until eventually it topples of its own weight.

Both parents incubate the eggs for more than a month. The young, born blind and helpless, do not fully mature until they reach four or five years old.

Size: 40"L, 80"W
Breeding range: Formerly throughout North America; today primarily in Alaska and western Canada, the Great Lakes area, Chesapeake Bay, and southern Florida

Winter range: Southern Canada west to coastal Alaska and south to southern California and Florida
Preferred habitat: Near large bodies of water
Preferred nest site: In a fork near the top of a giant tree, usually 50 or 60 feet up

Clutch size: 1 to 3, usually 2
Incubation period: 35 days
Nestling period: 72 to 74 days
Broods per season: 1
Preferred food: Fish, small to medium-size mammals

Red-Tailed Hawk

(*Buteo jamaicensis*)

SOARING above western prairie or eastern farmland, the redtail seems the epitome of freedom. Its tail is redder and broader than the tails of other hawks; its body is gray above and white below. It courts on the wing, the male calling to the female, and mates for life. Its nest is a casual collection of sticks, 30 to 40 feet up in a tree, a saguaro, or on a cliff.

In the fall red-tailed hawks fly south, floating on updrafts along north-south mountain ridges. Hundreds of them pass Pennsylvania's Hawk Mountain every autumn.

Size: 18"L, 56"W
Breeding range: Central Quebec west to Alaska and the Yukon, south to Mexico and Florida
Winter range: Central New England west to southwestern British Columbia, south

Preferred habitat: Dry wooded or open areas, marshes, deserts
Preferred nest site: Forest tree, isolated tree, cliff
Clutch size: 1 to 5, usually 2 or 3
Incubation period: 28 days
Nestling period: 4 to 5 weeks

Broods per season: 1
Preferred food: Small mammals, amphibians, reptiles, nestlings, insects

Accipitridae

Northern Harrier

(Circus cyaneus)

NORTHERN harriers engage in spectacular nuptial flights. The male dives repeatedly from a height of about 60 feet (occasionally from as high as 500 feet) to within 10 feet of the ground, then climbs and repeats the performance, usually about twenty-five times. The record observed is seventy-one uninterrupted dives and swoops.

The female northern harrier builds on or near the ground in freshwater or saltwater marshes or other wet places. The male gathers the material — sticks, straw, and dry grasses — for the 2- to 18-inch-high structure.

Size: 16½"L, 42"W
Breeding range: Most of North America south to northern Virginia and west to Texas panhandle, California

Winter range: Nova Scotia, southern New England west to the Pacific and south to South America
Preferred habitat: Marshes, fields
Preferred nest site: On or near the ground

Clutch size: 3 to 9, usually 5
Incubation period: 24 days
Nestling period: 5 to 6 weeks
Broods per season: 1
Preferred food: Small mammals and birds

145

Osprey

(Pandion haliaetus)

THIS magnificent bird is admired by dwellers along large rivers and both our coasts for its spectacular fishing technique and its huge, rubbish-filled nest. If an osprey, or fish hawk, nests nearby, you are fortunate — they are fascinating to watch. They dive from 100 feet up right into the water to seize their prey with their talons; some have even drowned clutching heavy fish. Several ospreys may even band together to battle in mid-air against a bald eagle trying to rob their catch.

Ospreys nest in loose colonies near fresh or salt water. The nests may be near the ground or up to 60 feet high on trees or poles. Often reused year after year, they are constructed of small to enormous sticks and lined with bark, sod, seaweed, and grass. New materials are added every spring. Grackles have been known to nest in the lower perimeter of osprey nests, dining on fallen scraps.

Both parents defend the nest, but the female does most of the incubating. The young are born blind and helpless.

Size: 22"L, 54"W
Breeding range: Practically worldwide. In North America: Alaska, Canada south along coast to Florida, along Gulf coast to Louisiana; in west, south to northern California, New Mexico

Winter range: Florida and the Gulf states and south, Baja California, southwest Arizona, Mexican coastlines
Preferred habitat: Near fresh or salt water
Preferred nest site: From ground level to as high as 60 feet up, always above surrounding area

Clutch size: 2 to 4, usually 3
Incubation period: 28 to 35 days
Nestling period: 8 to 10 weeks
Broods per season: 1
Preferred food: Fish

Falconidae

American Kestrel

(Falco sparverius)

AMERICAN kestrels claim a territory of 250 acres and defend it from March until June. Male American kestrels dive from the heights and swoop upward over the nesting area. American kestrels have an excited *klee-call* when disturbed, a *whine-call* when transferring food, and a *chitter-call* during approaches and copulation. Kestrels court for about six weeks, copulating as often as fifteen times a day, then separating and preening.

Kestrels, which accept nesting boxes, are cavity nesters, using tree holes 15 to 30 feet from the ground with an opening 2 to 4 inches in diameter. They do not line the cavity. Flickers and squirrels compete with them for nesting holes.

American kestrels brood their young for thirty days; both parents have brood patches. The young hatch over a period of three to four days. By the ninth day, the female broods only at night. Until that time, the male brings all the food, but then the female largely takes over the job. The fledgling period lasts fourteen days, during which the youngsters are active but still beg from the parents.

Size: 11"L
Breeding range: Nova Scotia west to central Alaska and Northwest Territories, south to Baja California, east to Florida and the Gulf states, South America

Winter range: Southern Ontario west to British Columbia and south
Preferred habitat: Open areas, from forest edges to cities
Preferred nest site: Cavity in tree, buildings, nestbox
Clutch size: 3 to 7, usually 4 or 5

Incubation period: 29 to 31 days
Nestling period: 30 to 31 days
Preferred food: Insects, small mammals, reptiles, amphibians, birds

GAME BIRDS —
GROUSE, QUAIL, AND PHEASANTS

THERE'S a bit of overlap here, as some water birds are considered game birds (ducks and geese, for instance). For the sake of simplicity, therefore, we'll start with game birds that live on land and are willing to visit feeding stations in selected locations.

"Selected locations" is the key phrase for all the birds in this chapter. Urban bird feeders aren't likely to see quail or pheasants. Suburban dwellers frequently can attract these birds but shouldn't expect grouse. Rural bird lovers may get grouse and more and more frequently can expect to see the once-reclusive wild turkeys in woods and fields.

All of these birds belong to the order Galliformes and are considered chickenlike birds. They live mostly on the ground and are primarily eaters of plant foods. They have strong legs and feet with which to scratch for food. They seldom migrate. In most cases the fancy-looking males keep harems but do none of the chores of parenting. Their young are precocial and able to run about as soon as they are hatched. The domestic chicken, to name the single most valuable bird, is a member of this order, and so is the turkey. In fact, the domestic turkey is the very same species as the wild one.

We'll begin with native birds, the Tetraonidae family of **grouse**. The grouse, often popularly called partridges, are northern birds that tend to be shy near settled areas. It's quite an accomplishment to entice them, and locale is extremely important.

Grouse exhibit interesting courtship behavior. During courtship, inflatable air sacs on their necks increase the resonance of their calls, which are referred to as *booming*. At their annual *booming grounds* (the site they select as a courting arena) as many as fifty cocks may be seen together, heads and necks thrust forward, with inflated air sacs, drooping wings, and raised tails. After booming, they stamp their feet, turn, and run in circles. Mating occurs on the spot after the ceremonial dances. Males are polygamous; the female departs the booming grounds and goes about her matronly duties alone.

The family Phasianidae is found nearly all over the world and includes **quail, pheasants, partridges,** and **peacocks**.

Left to right: Wild turkey hen, ring-necked pheasants, wild turkey cock, northern bobwhite (upper center), California quail (lower center), Gambel's quail, ruffed grouse.

Ruffed Grouse

(Bonasa umbellus)

THE ruffed grouse, found coast to coast in the North, is fond of goldenrod, clover, apples, cottonwood, birch, alder, blueberries, high-bush cranberries, dandelions, hazelnuts, chokecherries, and Japanese barberry. Norman Wight, an avid birder in Maine, scatters scratch feed and apples for them in the pine trees near his barn. Sometimes grouse visit a yard for oats. Ruffed grouse nest under logs or at the base of a tree in hollows they line with leaves or pine needles. During courtship the ruffed grouse makes a drumming sound with his wings.

Size: 14"L
Breeding range: Newfoundland and southern Labrador west to southern Manitoba, and Alaska, south to northern California, and south in the Appalachians to South Carolina

Winter range: Resident in breeding range
Preferred habitat: Woodlands with clearings and second growth
Preferred nest site: Under a log, at the base of a tree, in dense undergrowth

Clutch size: 8 to 15, usually 9 to 12
Incubation period: 21 to 24 days
Nestling period: Less than 1 day (precocial)
Broods per season: 1
Preferred food: Seeds, insects, fruits

Sharp-tailed Grouse
Spruce Grouse

(Tympanuchus phasianellus)

(Dendragapus canadensis)

THE sharp-tailed grouse, native to parts of Oregon, Utah, New Mexico, and Nebraska, is unlikely to come to your feeding station but may appear on your land if you grow aspen, birch, oaks, or cranberries. Gravel is essential for these birds. The spruce grouse, found in northern coniferous forests, will eat birch buds and catkins as well as the needles and buds of larch, pine and, of course, spruce.

When courting, the sharp-tailed grouse inflates purple sacs on its throat, shakes its tail, and utters a gurgling crow.

Sharp-tailed Grouse

Size: 15"L
Breeding territory: Alaska and Canada east to western Quebec; south to northern Wisconsin, Nebraska, and northeastern New Mexico; west to Idaho and Washington east of the Cascades
Winter range: Resident in breeding range
Preferred habitat: Open woodlands and brush
Preferred nest site: Depression in grass
Clutch size: 7 to 15
Incubation period: 21 days
Nestling period: Less than 1 days (precocial)
Broods per season: 1
Preferred food: Vegetation, insects

Spruce Grouse

Size: 13"L
Breeding range: Newfoundland and Labrador west to Alaska; south to northern Washington and western Montana, south-central Saskatchewan, northern Great Lakes region, and northern New England
Winter range: Resident in breeding range
Preferred habitat: Coniferous forests, especially spruce
Preferred nest site: On ground in brush piles or under a conifer with low branches
Clutch size: 4 to 10, usually 6 to 8
Incubation period: 22 to 25 days
Nestling period: Less than 1 day (precocial)
Broods per season: 1
Preferred foods: Buds and needles of conifers, berries, insects

Northern Bobwhite

(Colinus virginianus)

PROBABLY the best known quail is the northern bobwhite. Found naturally from the Rockies east to the Atlantic, except in the extreme Northeast, it has been introduced to the Pacific Northwest, Idaho, Hawaii, and west Texas. It's the smallest of our native quail except for the Harlequin quail, which is found only in a limited range. The bobwhite is inconspicuous looking but a friendly sort. When there still remained vacant lots around my parents' home in central Florida, they hosted a visiting bevy of quail several times each day. In season, the bobwhites were followed by their young. Mum and Dad's Scottie, Patti, loved to watch them.

On several occasions, a juvenile bird detached itself from the flock to investigate Patti carefully. The dog always endured this examination courteously, permitting the little one to rejoin the others in peace.

How do you attract these amiable birds to your yard? Their preferred habitat is brush. They eat seeds and some insects, such as beetles, crickets, and spiders. They'll eat grain you scatter, especially oats and cracked corn. Plants that attract them include sand cherry, redbud, bittersweet, white ash, American beauty berry, American beech, blackberries, vetch, clover, lespedeza, alfalfa, mulberries, amaranth, four o'clocks, and locust.

The nest of the northern bobwhite is built by both parents in a hollow in the grass and is lined with grasses. They weave weeds into an arch over the nest.

Size: 8"L
Breeding range: Major range southwestern Maine west to South Dakota, south to the Gulf of Mexico and east to the Atlantic; also eastern Washington, Wyoming, eastern Colorado, eastern New Mexico, eastern Mexico

Winter range: Resident in breeding range
Preferred habitat: Brush and farming country
Preferred nest site: A hollow in the grass
Clutch size: 12 to 20, usually 14 to 16

Incubation period: 23 to 24 days
Nestling period: 1 day or less (precocial)
Broods per season: Usually 2
Preferred food: Vegetation, seeds, insects

Gambel's Quail
California Quail

(Callipepla gambelii)

(Callipepla californica)

Gambel's quail

California quail

GAMBEL'S quail, a bird of the Southwest, is also called the desert quail. It is a stunningly beautiful bird, a little larger than the bobwhite, with a black plume extending forward from its crown. Strange and wonderful, it is reminiscent of the exotic appeal of the peacock.

In California, Oregon, and Nevada is found a similar bird, the California quail. Both of these species are able to go without water if sufficient succulent vegetation is available. They eat insects as well as grains, fruits, and berries.

Gambel's quail

Size: 8½"L
Breeding range: Deserts of southern Nevada and Utah, central Idaho, western Colorado, southeastern California, Arizona, New Mexico, west Texas, northwest Mexico
Winter range: Permanent resident in breeding range
Preferred habitat: Thickets near water
Preferred nest site: On ground
Clutch size: 10 to 16
Incubation period: 21 to 24 days
Nestling period: 1 day or less (precocial)
Broods per season: 2
Preferred food: Insects, grains, berries, fruits

California quail

Size: 8"L
Breeding range: Southern British Columbia to Oregon, California, Baja California; also northern Nevada, central Utah, western Idaho
Winter range: Permanent resident in breeding range
Preferred habitat: Edges of woods, brush, parks, farms
Preferred nest site: On ground
Clutch size: 10 to 17
Incubation period: Unknown
Nestling period: Less than 1 day (precocial)
Broods per season: 1
Preferred food: Insects, grains, berries, fruits

Ring-Necked Pheasant

(*Phasianus colchicus*)

THE ring-necked pheasant is a large and handsome bird native to Asia. It has successfully adapted, especially to grain farming areas, but is not found in the South or Texas. Many states stock pheasants for hunting. If these birds visit your property, they'll readily accept cracked corn, and they appreciate a supply of grit when there is snow cover. They may wander into your garden for asparagus seeds, and they like dandelions and chokecherries. They're often seen foraging in fields for waste grain. Usually, a cock supervises a number of hens, and fights both other males and domestic roosters.

Ring-necked pheasants nest on the ground in a natural hollow lined with weeds, grasses, and leaves. The hen does all of the work of incubating and parenting, usually in the company of other hens.

Size: 27"L
Breeding range: Southern New England west to British Columbia, south to southern California, ranging north from New Mexico and east; primarily north of the Mason-Dixon line; stocked in many areas as game bird

Winter range: Resident in breeding range
Preferred habitat: Cultivated fields, brushy edges
Preferred nest site: On the ground in a natural hollow, in grass or weeds
Clutch size: 6 to 15, usually 10 to 12

Incubation period: 23 to 25 days
Nestling period: Less than 1 day (precocial)
Broods per season: 1 or 2
Preferred food: Grains, seeds, vegetation

Phasianidae

Wild Turkey

(Meleagris gallopavo)

THIS noble bird had disappeared from much of its native range until recent years, when widespread game management efforts successfully reestablished it through much of the Northeast. Its range covers Texas and Oklahoma east to Florida and well into New England. Thus it isn't as rare to see a wild turkey in many parts of the country as it was twenty years ago. Preferring woods to open land, wild turkeys often forage in a sociable group looking for acorns, aster seeds and leaves, sumac fruits, insects, and the berrylike cones of juniper bushes. They roost in the tops of tall trees, preferably pines.

Like many members of the order Galliformes, male turkeys maintain a harem — often comprising fifteen hens — and don't help with nesting or incubating. Wild turkey males even often interbreed with domestic turkeys. To make up for the lack of paternal help, turkey hens often help each other raise their broods. They nest in a well-hidden depression in dead leaves on dry ground, laying ten to fourteen pale tan eggs and brooding them for twenty-eight days. When they hatch, the chicks don't eat for two days, but they're able to run around.

Size: 34"L
Breeding range: Central and southern New England, south to Florida, the Gulf states, and Mexico; north to Colorado and east; reintroduced in much of former range
Winter range: Resident in breeding range

Preferred habitat: Open woodlands in hilly or mountainous areas, stands of pines, farmlands
Preferred nest site: Depression in dry ground
Clutch size: 8 to 20, usually 10
Incubation period: 28 days

Nestling period: 1 day (precocial)
Broods per season: 1
Preferred food: Acorns, berries, plants, insects, corn, apples

PIGEONS AND DOVES

An easy-to-attract family is the Columbidae, pigeons and doves, in the order Columbiformes. We ought to first clear up the sticky business of the difference between pigeons and doves. Simple: There is none. The terms can be (and are) used interchangeably.

There's a popular misconception that doves, symbols of peace and love, are gentle. Don't you believe it. Tender to their mates (though occasionally inconstant), protective of their offspring, yes. But two males meeting when the breeding season gets under way means a horrendous fracas, with the two beating one another with their wings. Wild or feral losers usually fly away unharmed; domestic or captive ones are often seriously (even fatally) injured. Courtship follows these furious battles.

Under the feeder the dominant doves relentlessly chase others — of their own or any other ground-feeding species — away from choice morsels. Those soft, cooing noises have misled us all.

Now for the wonderful peculiarities of pigeons and doves. Unlike other birds, they don't need to raise their heads to drink; they just submerge their beaks up to the nostrils and sort of inhale. Normally, they drink mornings and evenings, and since they're strong, swift flyers, they may go rather long distances for water. And while they eat mostly grains, seeds, and fruit, they seem to require considerable salt and often can be seen scavenging for both salt and gravel (for their crops) along roadsides.

Doves are careless nest builders, whether they choose a manmade structure, a tree, or the ground for the nesting site. Egg casualties, from accidents or flimsy nest constructions, are commonplace. Both parents share the incubating chores and the care of the young. Both also produce the secretion known as pigeon milk that feeds the nestlings early in life. Since neither sex shows any interest in housekeeping tasks, the nest and its environs are definitely untidy — one might say filthy — by the time the squabs are ready to leave home.

Pigeons and doves are found all over the world, both in tropical and temperate climates, and object not in the least to encroaching civilization. The passenger pigeon, it is true, is extinct, though it was once probably our most numerous native species. Its extinction is a reprehensible example of squandering a resource, of annihilating a species that seemed inexhaustible. The young birds were not only easier to take but also commanded a higher price at the market than older ones. Their slaughter and the accompanying disturbance to flocks, combined with the usual run of natural catastrophes, had disastrous consequences.

Left to right: Inca dove, rock dove, mourning dove, ground dove.

Rock Dove

(Columba livia)

WHAT we have now, in tremendous numbers, is the rock dove, the commonplace pigeon of urban areas. Like starlings and house sparrows, it was introduced from Europe. It is abundant in wild, feral, and domestic conditions. The wild variety is distinctive, but the feral and domestic birds come in all sorts of color combinations. You're familiar with rock doves; you know one when you see it, whatever the markings may be. Their diet consists entirely of plant food. They'll eat grains of all kinds and peanuts.

The rock dove defends a very small territory — only the immediate vicinity of the nest itself. During the breeding cycle, the male will bow at opponents and strike at them with an outstretched wing. Both rock doves and mourning doves mate for life. With puffed feathers and trembly wings, the males strut, bow their heads, and coo.

Rock doves nest in colonies or singly, on ledges, bridges, or barns. The nest is a shallow, flimsy platform carelessly constructed of short twigs, grasses, straw, and debris of various sorts. The outside diameter is 8 inches. The chicks stay in the nest about ten days, fed on pigeon's milk and regurgitation and eventually insects and fruit. During the fledgling period, they get little or no food from the parents; though they beg, the results are minimal.

Size: 13"L
Breeding range: Throughout temperate North America
Winter range: Permanent resident throughout its range
Preferred habitat: Near human habitation
Preferred nest site: On ledges, bridges, barns, singly or in colonies
Clutch size: 1 to 2, usually 2
Incubation period: 17 to 19 days
Nestling period: 21 to 28 days
Broods per season: 2, 3, or more
Preferred food: Seeds, grains, handouts

Mourning Dove

(Zenaidura macroura)

THE settlement of the continent has provided favorable habitat for the mourning dove, and it has become more abundant throughout its range. In some areas it is classified as a songbird and is, therefore, protected; in others it is considered a game bird.

Mourning doves are said to be especially fond of hemp and millet seeds. We see them daily in pastures and mowings, eating weed seeds. They also partake of beechnuts, small acorns, and grain. Most of the grain they eat is from the ground, so they aren't normally considered a menace to agriculture.

Formerly known as the Carolina or turtle dove, the mourning dove mates for life. When courting, the males perform the "tower dance," in which they fly to a height of about 30 feet and flutter, descend a little and flutter again,

and so on. Sometimes the male circles above his mate, his tail fanned. As he spreads his tail so that the white shows, he calls, closes his tail, waits, and repeats the entire procedure.

Mourning doves will use a cone-shaped nesting platform of ¼-inch or ⅜-inch hardware cloth placed in forked branches, 6 to 16 feet from the ground. Otherwise, the female builds a frail platform of twigs, which the male collects, on a horizontal branch 3 to 20 feet above the ground. It may be lined with small twigs, pine needles, and grasses. Sometimes the female simply lays her eggs on the ground instead of in a nest; sometimes she uses an old robin, grackle, or blue jay nest. Whatever the

nest, mourning doves don't clean it, and it quickly becomes a mess.

Mourning doves are devoted parents that brood the young — which are blind, naked, and helpless at hatching — until they are fledged at thirteen to fourteen days. The male broods eight hours a day; the female, the rest of the time. The young are fed on pigeon's milk by both parents; they close their beaks on that of the nestling and "pump" the secretion into the young birds for a period of fifteen to sixty seconds. Later on they feed seeds, worms, and insects.

Size: 12"L
Breeding range: Southern Canada west to British Columbia, south to the Bahamas and Mexico
Winter range: Southern Maine west to extreme southwest British Columbia and south to Central America
Preferred habitat: Open woodland, agricultural, and residential areas
Preferred nest site: Horizontal branch of conifer, from 10 to 50 feet high
Clutch size: 2 to 4, usually 2
Incubation period: 13 to 14 days
Nestling period: 12 to 14 days
Broods per season: 2 in North, 3 or 4 (occasionally to 6) in South
Preferred food: Seeds, grains

Ground Dove

(Columbina passerina)

THE ground dove, a bird of the southern United States, is much smaller than those so far mentioned, only somewhat larger than a sparrow. It's mostly gray, with stubby black tail and chestnut wing linings.

Expect it to come to feeders for small grains, cracked corn, or scratch feed and small seeds. It regularly visits lawns, gardens, and farms but is considered shy. The ground dove eats a few insects, but most of its diet consists of weed seeds, waste grains, and berries.

Ground doves are even dirtier nesters than mourning doves, especially when they use old mourning dove nests. Sometimes they build a loose foundation of twigs or pine needles lined with rootlets and grasses placed in a vine or bush or on a stump, fence post, or horizontal branch. Other times they nest on the ground in a slight depression, which is lined with grasses in a haphazard manner, if at all. Whatever they use, on the ground or up to 25 feet above it, the same nest serves for all the broods of a given season. They brood their young for fourteen to sixteen days.

Size: 6½"L
Breeding range: South Carolina west to California and south to the Florida Keys and Central America
Winter range: Primarily a resident, but withdraws from northernmost part of range

Preferred habitat: Open woodlands, agricultural and residential areas
Preferred nest site: From ground level to as high as 25 feet, on a horizontal branch
Clutch size: 2
Incubation period: 12 to 14 days

Nestling period: 14 to 16 days
Broods per season: 2, 3, or more
Preferred food: Seeds, grains, berries

Inca Dove (*Scardafella inca*)

THE Inca dove, larger than the ground dove, has a long, white-edged tail. It's found in the Southwest, in open lands, parks, farms, lawns, and gardens. It often forages for grain with domestic chickens and other livestock. It also eats weed seeds.

Among Inca doves, the older birds pair before the younger ones do. Both males and females engage in much head bobbing, cooing, and mutual preening; the males add vertical tail fanning to their performance. Inca doves *coo* while head bobbing; the male song is *cut-cut-ca-doah*. After courtship and before the laying of eggs, the male Inca dove defends a cleared area with trees nearby, from 30 x 50 to 70 x 100 yards. At other times, Inca doves are highly social.

Inca doves spend about three days building a loose, shallow, cup-shaped nest of twigs, grasses, rootlets, and plant fibers. The male collects the nesting materials and passes them to the female, who puts them in place. The nest is unlined, and may be from 5 to 25 feet from the ground, in a bush, vine, or tree, on a horizontal fork. The female shapes the interior of the nest with her body.

Inca doves brood their young for fourteen days, the female from late afternoon until morning, the male from morning to late afternoon. Once the fledglings leave the nest, their parents feed them for a week and then start another brood; the juveniles join a flock of other juveniles.

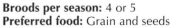

Size: 8"L
Breeding range: Extreme southwestern United States
Winter range: Resident of breeding range
Preferred habitat: Open land in rural and suburban areas

Preferred nest site: Horizontal branch or fork of tree, from 4 to 25 feet high
Clutch size: 2
Incubation period: 14 days
Nestling period: 14 days

Broods per season: 4 or 5
Preferred food: Grain and seeds

159

OWLS

WITH their huge, round, yellow eyes, haunting calls, and nearly silent night flight, owls command our awe as do few other birds. We are lucky indeed if we have them in our neighborhoods, for they keep the population of rats and mice in check.

Left to right: Great horned owl, eastern screech owl, northern saw-whet owl, common barn owl.

BARN OWLS

Tytonidae

Common Barn Owl

(Tyto alba)

RANGING through all but the northernmost tier of the United States, the barn owl is extremely important as a predator of rats and mice and should be protected.

The barn owl is pale in color, and its face has been said to resemble a great white heart. Its wingspan can reach nearly four feet and females are often larger than males. Males, females, and juveniles have similar coloring.

Barn owls don't build nests; the female lays her eggs on owl castings (disgorged encapsulated fecal material) in cavities, usually in barns, silos, or steeples. She incubates the eggs for up to 34 days.

Size: 14"L, 44"W
Breeding range: U.S. except Montana, North Dakota, Minnesota, northern Michigan, Maine
Winter range: Resident in most of breeding range but retreats from northern parts

Preferred habitat: Woodlands, farmlands, often near human habitation
Preferred nest site: Barns, buildings, birdhouses
Clutch size: 3 to 11, usually 5 to 7

Incubation period: 21 to 24 days or 32 to 34 days (depending on source of information)
Nestling period: 59 to 60 days
Broods per season: 1 or 2
Preferred food: Mice

Eastern Screech Owl

(Otus asio)

THE first time you hear this aptly named owl's wailing call it can give you a shiver. The eastern screech owl is eight inches long, about half the length of the barn owl, with a wingspan of about two feet. It can be streaked with gray or red and has feathery tufted "ears" that it erects when excited. Of course, as with the great horned owl, its true ears are hidden beneath feathers on the side of its head.

Screech owls are cavity nesters and are less consistent in their nesting habits than barn owls. They lay eggs on leaves, rubble, or whatever happens to be in the cavity; they accept nestboxes in trees, especially in orchards.

Insects, mice, birds, lizards, crawfish, spiders, worms, and frogs are all on the screech owl's menu.

Size: 8"L, 22"W
Breeding range: Maine, Quebec west to Montana, Wyoming, south to eastern Mexico and Florida
Winter range: Resident in breeding range

Preferred habitat: Open woodlands to shade trees
Preferred nest site: Natural cavities or abandoned woodpecker holes
Clutch size: 2 to 7, typically 4 or 5

Incubation period: 21 to 30 days
Nestling period: 30 days
Broods per season: 1
Preferred food: Small rodents and insects

161

Great Horned Owl

(*Bubo virginianus*)

RANGING from deep woods to desert, the great horned owl is large and fierce. The female's wingspan can reach five feet; the male's is just a few inches shorter. Plumage can be very dark or very pale in color, depending on its habitat, with conspicuous ear-tufts, barring across its breast, and powerful talons. Great horned owls swoop on silent wings after prey ranging in size from rabbits and turkeys to insects.

The pair nest in caves, gorges, deep woods, or in the abandoned nest of a crow, and they share the job of brooding their one to three eggs for up to five weeks.

Size: 20"L, 55"W
Breeding range: Arctic south to Strait of Magellan
Winter range: Permanent resident in breeding range
Preferred habitat: Deep woods to city parks to open fields

Preferred nest site: Old nests of large birds, cavities in trees
Clutch size: 1 to 3, usually 2
Incubation period: 28 to 35 days
Nestling period: 40 to 45 days
Broods per season: 1

Preferred food: Small mammals, birds, reptiles

Strigidae

Northern Saw-Whet Owl (*Aegolius acadicus*)

THIS shy, small owl resembles a screech owl without ear-tufts. The smallest of our nocturnal raptors at eight inches in length, it often nests in an abandoned woodpecker or flicker hole.

The female takes charge of incubation duties for up to four weeks. The nestlings fledge by autumn and can then begin to learn to hunt. Mice and insects are this owl's foods of choice.

The name "saw-whet" describes its call, which sounds like a saw being sharpened.

Size: 8"L, 17"W
Breeding range: Disjunct breeding range: western U.S., southern British Columbia and southeastern Alaska; northeastern U.S., Appalachian Mountains, and southeastern Canada

Winter range: Southern California, Great Plains, Missouri, and Arkansas to South Carolina and Maryland
Preferred habitat: Forests, groves, thickets
Preferred nest site: Tree cavities, nestboxes

Clutch size: 4 to 7
Incubation period: 21 to 28 days
Nestling period: 27 to 34 days
Broods per season: 1
Preferred food: Small mammals, insects

SWIFTS AND HUMMINGBIRDS

SWIFTS and hummingbirds, which make up the order Apodiformes, live their lives mostly on the wing. Indeed, the common swift of the Old World eats, drinks, mates, and even sleeps while flying.

Found only in North, Central, and South America, the hummingbird family, the Trochilidae, contains more than 300 species, ranging in size from 2¼ inches to 8½ inches. Many of these species have very distinct ranges — the streamertail, for example, lives only on the island of Jamaica. Only seventeen species enter the United States, and neither the largest hummingbird (a native of the Andes)

nor the smallest (a resident of Cuba) is among them. Even seventeen strikes me as an embarrassment of riches, since I live in an area that boasts only one: the ruby-throated hummingbird.

As a family, hummingbirds are remarkable, no two ways about it. For one thing, they're tiny. That Cuban hummer is the smallest bird in the world. Here in the U.S., our smallest, the calliope, is under three inches, and our largest barely makes five inches. Most of them measure between three and four inches.

Size isn't their only claim on our attention. Take the matter of wings. Most birds

have powered downstrokes. In hummingbirds both upstroke and downstroke are powered. They can fly backward, forward, and sideways, or, like a helicopter, they can hover. They have the fastest wingbeat of all the birds, and the name "hummingbird" derives from the sound made by their wings.

They also have the most rapid heartbeat. To maintain themselves during the daytime, they need to eat every ten to fifteen minutes. (They have the highest relative food consumption of any bird known.) At night, hummingbirds fall into a torpor resembling hibernation, a condition requiring only a twentieth of

Clockwise from upper left: Ruby-throated hummingbird, Anna's hummingbird, chimney swift, rufous hummingbird, Allen's hummingbird, black-chinned hummingbird pair.

their normal daytime energy output, and their body temperature drops almost to that of air. Next morning they arouse almost instantaneously.

Hummingbirds can perch, but they have such weak feet and legs that they can scarcely walk. Their plumage is typically brilliant, iridescent. Their beaks are very long and needlelike, and their tongues are long, double tubed, and sticky, enabling them to cope with sap, nectar, small insects, and hummingbird feeders.

All of our hummingbirds migrate, too. The tiny **ruby-throated hummingbird** flies across the Gulf of Mexico to and from the Yucatán at an estimated 30 mph. The sexes migrate separately, the male arriving first, and move northward as flowers appear. They fly about 25 feet above land or water.

Like swifts, hummers conduct their courtship in the air. Despite all their complicated courtship activity, once they mate the male disappears completely, leaving all the household chores to the female.

Hummingbirds build cup-shaped nests of great beauty and delicacy. Like swifts, hummingbirds use a saliva-like glue to attach their nests to the chosen site. Their nests are lined with plant down, feathers, or fur. About an inch deep and an inch in diameter, from a distance they could be mistaken for a knot or a gall on a tree.

After mating, the male leaves for other adventures and the female builds the nest, incubates the two pea-size eggs, and cares for the nestlings. They're fed at first by regurgitation, as often as five times an hour. The placing of that needle-type bill into the nestling's gape has been likened to a surgical procedure.

Hummingbirds spend the time left over from feeding and reproducing in quarreling — with each other (assuming two birds of the same sex) or anybody else that comes along. They're unimpressed by size as a deterring factor and are just as willing to tangle with such enemies as crows, kingbirds, and hawks as with one of their own species. They can escape most enemies because of their superior maneuvering abilities and because they're capable of traveling sixty miles per hour, the speed of a hawk. Though many species of hummingbirds live in equatorial jungles, they may also be found in forests, gardens, deserts, plains, canyons, and mountains.

How do you attract these fascinating birds to your garden? With flowers, primarily. Use feeders, by all means, but not feeders alone. Someone gave one of our friends a hummingbird feeder because his patio was practically swarming with hummers. He swears that since he installed it his only visitors have been doves. It's my conviction that feeders are second in importance only to a supply of *natural* food. That means lots of flowers, which can only be considered an additional ornament to your property.

Experimental studies have convinced researchers that hummingbirds are first attracted to red flowers, but contrast is important. Next on their preferred list is white, followed by violet, orange, blue, yellow, and green. It has been suggested that green may be more attractive to them in arid country than it is in more humid areas.

Some flowers seem almost irresistible to hummingbirds. Particularly recommended are hibiscus, petunias, mimosa, azaleas, trumpet vine, lupines, tree tobacco, salvia, larkspur, flowering quince, monarda (bee balm), ajuga, buckeye, chinaberry, impatiens, zinnias, columbines, snapdragons, dahlias, gladiolas, and fuchsia. Flowers, particularly brilliant ones, attract hummingbirds. Who could ask for a better combination?

Chimney Swift

(*Chaetura pelagica*)

THE long-winged, short-tailed chimney swifts defend no territory whatever. Their group flights, trio flights, and V-gliding (groups of birds gliding in V-formation) are probably part of their courtship rituals. They mate in flight and at the nest site.

They are partial to dark areas, especially chimneys. The nest — a frail, thin, half-saucer 4 inches in diameter and attached to the inside of a structure — is built of twigs gathered in flight, broken off trees, by both parents. The twigs are held together by a glutinous secretion from the bird's mouth, popularly called "saliva," though it isn't, of course. It hardens and binds the twigs together and attaches the nest to the wall. Chimney swifts will use old nests, repaired. The female puts the finishing touches to the nest while incubating.

Both chimney swift parents brood the young during the nestling period, which lasts fourteen to nineteen days. Both also feed the young, first on regurgitated insects and later on pellets of insects bound together by the salivalike substance secreted by the parents' mouths. During the fledgling period of fourteen to eighteen days, the young are still dependent on their parents for food and stay near the nest. Eventually they begin to fly out and catch their own meals.

Size: 5"L, 12½"W
Breeding range: Eastern North America, from southern Canada south to Florida and the Gulf states
Winter range: Peru

Preferred habitat: The sky, near human habitation
Preferred nest site: Chimneys
Clutch size: 3 to 6, usually 4 or 5
Incubation period: 14 to 19 days

Broods per season: 1
Preferred food: Flying insects

Ruby-Throated Hummingbird

(*Archilochus colubris*)

NOW just a bit of more specific information about some of the hummingbirds you might expect to see. From a line extending through North Dakota and eastern Kansas south to eastern Texas, the area stretching to the Atlantic and Gulf Coasts is unlikely to host any but ruby-throated hummingbirds. In our garden, we usually see the first ones of the season feeding on the flowering quince. Quarrelsome little birds, 3 to 3¾ inches in length, they mutter irritably to themselves most of the time between sips. In addition to nectar, they ingest small insects and sometimes snack on maple or other sap seeping from bark punctured by sapsuckers.

When courting, the male ruby-throated hummingbird perches and waits for the female and then commences a pendulumlike courtship flight of from 3 to 40 feet, lasting anywhere from two to twenty minutes. Sometimes he rises high, dives, faces the female, and spreads wings and tail before rising again. In another ceremony, the male faces the nest and makes a sideways flight of about 25 feet east to west and then back again, stopping about every 3 feet to hover and hum. Males and females perform one characteristic courtship flight together. Starting about 2 feet apart, they fly up and down at a range of 10 feet for two or three minutes at a time in op-

posite directions so that when the male is at the top of the flight, the female is at the bottom, and vice versa.

The nest of the ruby-throated hummingbird is an inch deep and an inch in diameter (about the size of an eyecup) and is built from 5 to 50 feet high on a small limb of a tree in orchards or woods, often over water. The nest is made of bud scales covered with lichens and is lined with plant down, the whole held together by spider webs or caterpillar silk. Old nests are renewed the following season. Other hummingbirds build similar nests.

Size: 3"L
Breeding range: Nova Scotia west to southern Alberta, south to Gulf states and Florida
Winter range: Mexico, Central America

Preferred habitat: Mixed woodlands, gardens
Preferred nest site: Limb of tree
Clutch size: 2
Incubation period: 11 to 16 days
Nestling period: 14 to 31 days

Broods per season: 1 or 2
Preferred food: Nectar, small insects, sap

Black-Chinned Hummingbird   *(Archilochus alexandri)*

THE black-chinned hummingbird is a close relative of the ruby-throated (the females are virtually indistinguishable in the field) but is found only during migration in any of the same range. The black-chinned lives in semiarid country, chiefly from western Montana south to central Texas. It sometimes waits on a perch for flying insects, which it pursues in much the same way a flycatcher does.

The black-chinned hummingbird likes a nesting site near water, from 4 to 30 feet from the ground in a willow, sycamore, or (in arid country) cottonwood tree. The nest, made of plant down usually without lichen decoration, resembles a tiny yellow sponge. Black-chinned hummingbird females brood for twenty-one days, feeding the nestlings tiny insects.

Size: Between 3 and 3¾"L
Breeding range: West of the Rocky Mountains, southern British Columbia and northwestern Montana south to Texas, and lower California and northern Mexico
Winter range: Mexico

Preferred habitat: Semiarid country near water to semiwooded to suburbs
Preferred nest site: Shrub or tree near water
Clutch size: 2
Incubation period: 13 to 16 days

Nestling period: 21 days
Broods per season: 2 or 3
Preferred food: Nectar, small insects

168

Anna's Hummingbird

(Calypte anna)

ANNA'S hummingbird breeds in California and is larger than other hummers found there. Eucalyptus, orange, and red-flowering gooseberry blossoms are among the favorite nectar sources for Anna's hummingbird, which is as common in urban areas as in canyons. The courtship flight of Anna's hummingbird is similar to that of Allen's hummingbird.

Male Anna's hummingbirds squeak and warble, in pursuit calling shrill chittering notes: *ztikl-ztikl-ztikl.* Both males and females make a noise that sounds like *chip.*

Anna's hummingbird uses lichens or mosses, cushioned with feathers or fur, in its nest, which is often placed on a branch near or over water, 2 to 30 feet from the ground.

The female continues building her nest as she incubates the eggs. She broods for twenty-one days; the nestlings are fed tiny insects.

Size: 3½"L
Breeding range: Southwest British Columbia, western Washington, Oregon, and California to southwestern Arizona and extreme northeastern Mexico

Winter range: Resident in breeding range but wanders to southeast Arizona, northwest Sonora
Preferred habitat: Brush, gardens
Preferred nest site: Branch of bush or tree, often near or over water

Clutch size: 2
Incubation period: 14 to 19 days
Nestling period: 21 days
Broods per season: 2
Preferred food: Small insects, nectar

Rufous Hummingbird

(Selasphorus rufus)

THE rufous hummingbird is found as far north as Alaska, east to western Montana and south to central California. A veritable tiger among hummingbirds, the rufous hummingbird is the most aggressive of a highly aggressive family. Both sexes defend territories, a practice common to all hummingbirds. The female defends the nest area; the male, the feeding territory.

He's especially attracted to red flowers, but he likes sap, too. He catches insects from the air and also plucks them from plants and flowers.

Despite their aggressiveness, rufous hummingbirds sometimes nest close together, 2 to 20 feet from the ground. Sometimes they build a new nest on top of an old one. Female rufous hummingbirds have the entire responsibility of building the nest, incubating the eggs, and brooding and feeding the young for a period of about twenty days.

Size: 3½"L
Breeding range: Southeastern Alaska to Alberta, south to western Montana and west to northwestern California
Winter range: Western and central Mexico

Preferred habitat: Forest edges, from mountain meadows to lowlands
Preferred nest site: Bush or tree
Clutch size: 2
Incubation period: 12 to 14 days

Nestling period: 20 days
Broods per season: 2
Preferred food: Nectar, insects

Allen's Hummingbird

(Selasphorus sasin)

ALLEN'S hummingbird is similar to the rufous hummingbird, to which it is closely related. It's found only along the Pacific coast from southern Oregon to the Los Angeles area during the breeding season. Insects and nectar comprise its diet.

When courting, the male Allen's hummingbird flies to an altitude of about 75 feet and then dives, swoops upward, and flies back and forth rapidly in a semicircular arc, squealing. At the top of the arc he spreads his tail, shakes his body, and dives again. He repeats this performance several times and then rises to 75 feet again and starts over. Allen's hummingbird buzzes, hums, and squeaks.

Allen's hummingbird usually nests at a height of 12 feet, but the nests have been found anywhere from 1 to 90 feet from the ground.

Size: 3"L
Breeding range: Pacific Coast, southern Oregon to Ventura County, California
Winter range: Northwestern New Mexico

Preferred habitat: Wooded or brushy areas, parks, gardens
Preferred nest site: Bush or tree, sometimes in colonies
Clutch size: 2

Incubation period: 14 to 15 days
Nestling period: 21 days
Broods per season: 2
Preferred food: Nectar, small insects

Other Hummingbirds

THE smallest hummingbird usually found in the United States is the **calliope,** measuring 2¾ to 3½ inches. This mountain resident, which weighs a *tenth* of an ounce, breeds in the Cascades, Sierras, and Rockies.

Other hummingbirds found in the United States in very limited ranges include the **magnificent** (formerly Rivoli's) hummingbird (at five inches, its only rival in size in our country is the **blue-throated hummingbird**), the **buff-bel-**lied, **violet-crowned, white-eared,** and **Lucifer** hummingbirds. In the field, it is especially difficult to distinguish female and juvenile hummingbirds of different species.

WOODPECKERS

THOUGH the majority of birds at the feeder will be the perching birds — the order Passeriformes — there's no trick to attracting woodpeckers, which belong to the order Piciformes, family Picidae. Many of them are resident year-round in their ranges, although they may wander considerably after the breeding season is over.

Despite great differences in appearance, all woodpeckers have certain physical characteristics in common. Essential are heavy, chisel-like bills to extract the grubs and larvae of insects found under the bark of trees; a strong skull, which their hammering on the wood to get to these delicacies necessitates; and long, extensible tongues, the better to ingest their food once it's uncovered. Their legs and feet are sturdy, and most of the family have two toes pointing forward and two pointing backward, enabling them to cling to tree trunks with ease and their stiff tail feathers

act as a prop to hold them in an appropriate position as they hammer away at the trees.

The speed and skill with which woodpeckers can make holes in trees — or telephone poles or houses, for that matter — have often caused them to be viewed with a certain amount of alarm. When we built a house some years ago in a wooded area, we stumbled into a running feud with a hairy woodpecker that seemed intent on destroying the gable on the eastern side of it. I spent considerable time rushing outdoors to yell at him. There *can't* have been any insects under those shingles, so we surmised that we had removed some of his favorite feeding sites and he couldn't break himself of the location habit. Whether the suet we supplied helped convince him to stop devastating the house is a moot point. Orchardists have often been dismayed by the holes excavated

by woodpeckers, but the fact is that the holes are less likely to damage the trees than the wood-boring creatures that the woodpeckers search out. If other insects later take up residence in the cavities, woodpeckers are quite ready to dine on them in their turn.

Members of the woodpecker family are largely arboreal (living in trees), with the prominent exception of the flicker. They are cavity nesters, using the chips of wood they remove as nesting material, and some will accept a nesting box, especially if it's covered with bark. They definitely prefer a rustic decor.

Both sexes incubate the eggs, which are white. Nestlings often make a humming noise. They remain in the nest for a longer time than terrestrial birds do and are well enough developed to fly immediately after quitting the nest. Many woodpeckers roost in their nesting sites.

Left to right: Pileated woodpecker, yellow-bellied sapsucker, northern flicker, red-bellied woodpecker, red-headed woodpecker, Lewis's woodpecker, hairy woodpecker, downy woodpecker.

Northern Flicker

(Colaptes auratus)

"HEIGH-HO." "Yarrup." "Clape." "Hairy wicket." "Yellow hammer." That's just a sampling of the 125 names by which the flicker has been known. And there's no question that the flicker is known, all over the continent. For one thing, flickers are big — about thirteen inches (robins measure ten inches from bill tip to tail tip, blue jays twelve inches). They're noisy, too. The yellow-shafted flicker is found east of the Rockies; the red-shafted flicker is found from the eastern foothills of the Rockies to the Pacific. The red and the yellow are listed as two different species, but ornithologists have begun to think maybe they are actually subspecies. The two interbreed wherever their ranges overlap, and their observable differences are in plumage only.

Although northern flickers are present throughout much of their range during the entire year, they tend to shift their populations southward for the winter. Their territory is an area of half a square mile, but they defend only the nest area, displaying head bobbing, a frozen pose, and drumming (beating a tattoo with their beaks on a resonant branch). Courtship among flickers is simple but a little peculiar, consisting largely of head bobbing and a fluttering flight. The unusual part is that two females may court one male (he's the one with the black mustache, obviously).

They nest in tree cavities or telephone poles, but they're willing to use nesting boxes, preferably bark covered, mounted on poles, and facing south, with entrance holes three inches in diameter. Put a layer of wood chips or shavings on the floor to serve as nesting material. Normally they nest 2 to 90 feet from the ground in a cavity 7 to 8 inches in diameter with a 3-inch south-facing entrance hole.

The parents raise one or two broods a year. Like other woodpeckers, flicker nestlings make a humming noise if disturbed, which some observers think sounds like a swarm of bees. The nestlings are fed by regurgitation. For the first ten days the parents eat the fecal sacs; after that they remove them, dropping them at a distance from the nest. Males brood the young at night; brooding lasts twenty-five to twenty-eight days. As fledglings the flickers may stay with their parents for two or three weeks, still getting food from them.

You'll find flickers as frequently as downy woodpeckers

Size: 10½"L
Breeding range: Entire continent except northernmost Alaska, Keewatin, and northern Quebec; also absent as breeder from central Texas and eastern Mexico

Winter range: Central New England west to southern British Columbia, south to Mexico and Florida
Preferred habitat: Woods to suburbs
Preferred nest site: Cavity in dead tree

Clutch size: 3 to 10, usually 6 to 8
Incubation period: 11 to 12 days
Nestling period: 23 days
Broods per season: 1
Preferred food: Ants, other insects, wild fruits

around the lawn and garden, but you'll find them also in the woods or on the edges of swamps, in brushlands, or open country. Unlike most other woodpeckers, they hunt some of their food on the ground. Over half of their diet is insects, especially ants. They eat more ants than any other bird known. But they'll chomp also on beetles, wasps, crickets, grasshoppers, and caterpillars. Every once in a while they even catch insects on the wing. Flickers eat a wide variety of plant material, too. They like plums, raspberries, blackberries, and chokecherries, poison ivy berries and sumac, dogwood, magnolias, oak, and mountain ash, euonymus, black gum, beech, Virginia creeper — you name it, flickers will sample the fruit or nuts. At the feeder they snack on suet and peanut butter.

WOODPECKERS   *Picidae*

Pileated Woodpecker   *(Dryocopus pileatus)*

AN extremely eye-catching woodpecker, which occasionally visits feeders for nutmeats and suet, is the pileated woodpecker. It's the size of a crow and has a bright red crest. Although pileated woodpeckers are habitually birds of the deep woods, they seem to be adapting to man's presence and now are found even in wooded suburbs. They're rather quiet birds, but can be startling, to say the least. I saw my first one from horseback; when it flew by, both horse and I jumped about three feet.

Pileated woodpeckers are known by some interesting names, including logcock, Wood Kate, and (a series I can only assume resulted from first impressions) Great God woodpecker, Good God woodpecker, and Lord God woodpecker. They usually nest in dead trees or tall stumps, in which they dig prodigious cavities. Their insect food includes carpenter ants and beetles in all stages of development. Seeking these creatures in tree trunks, they excavate holes 4 to 8 inches deep and rectangular in shape. They also eat wild fruits, berries, and nuts.

Size: 15"L
Breeding range: Nova Scotia to southeast Yukon, south to central California, east to Gulf states and Florida (absent from most of western U.S.)
Winter range: Permanent resident in breeding range

Preferred habitat: Deep woods to city parks
Preferred nest site: Tree cavity
Clutch size: 3 to 4
Incubation period: 18 days
Nestling period: 26 days
Broods per season: 1

Preferred food: Larvae and adults of carpenter ants and other insects, wild fruits, acorns, beechnuts

Red-Bellied Woodpecker (*Melanerpes carolinus*)

THE red-bellied woodpecker is also called the zebra woodpecker, the chad, and the ramshack. Although often thought of as a representative southern bird, it's found all over the United States east of the Plains, except New England north of Connecticut. It doesn't actually migrate, but it shifts southward after the breeding season and does a little wandering in the fall. Only a third of its diet is insects: ants, beetles, grasshoppers, and caterpillars. Its vegetable food includes cashews, beechnuts, almonds, butternuts, acorns, and various seeds and fruit, especially oranges. At the feeder, pieces of orange are particularly well received, but suet, peanut butter, nutmeats, and cracked corn are accepted, too.

Red-bellied woodpeckers build a gourd-shaped nest 1 to 18 inches deep in a dead tree, anywhere from 50 to 130 feet from the ground. It takes them a week to ten days to excavate the interior to their design. The entrance hole is 1¾ to 2 inches in diameter.

The parents brood the young for fourteen or fifteen days. Their nestlings eat insects at the age of four or five days, but it's likely they're fed by regurgitation before then. The nestlings sound like bees in the nest.

Size: 8½"L
Breeding range: New York, New Jersey, Pennsylvania west to southern Minnesota and southeastern South Dakota, south to central Texas, Gulf states, and Florida

Winter range: Permanent resident in all but northernmost portion of breeding range
Preferred habitat: Woodlands to towns
Preferred nest site: Tree cavity, nesting box

Clutch size: 3 to 8, usually 4 or 5
Incubation period: 14 days
Nestling period: Not known
Broods per season: 1 or 2
Preferred food: Insects, mast, corn, wild fruits

Red-Headed Woodpecker *(Melanerpes erythrocephalus)*

THE red-headed woodpecker is found in most of the same area as the red-bellied woodpecker, but also ranges up into northern New England. Its food choices are similar to those of the red-bellied woodpecker, and it exhibits the same tendency to store acorns and nuts in crevices for later use. Friends in Florida had a red-headed woodpecker visit their feeder regularly for sunflower seeds, which he carried to the eaves and dropped down a pipe supporting their ham radio antenna.

The red-headed woodpecker likes beechnuts, acorns, and cherries, chokecherries, other fruits and berries, and grain. It eats insects for about a third of its diet, sometimes catching flies on the wing. Peanut butter, corn, nutmeats, and suet attract this noisy, robin-size bird to the feeder.

Size: 7½"L
Breeding range: Southern New England west to southwestern Alberta and central Montana, south through eastern New Mexico to the Gulf states and Florida
Winter range: Migrates from northern and western parts of range, resident from Pennsylvania to southern Minnesota and south
Preferred habitat: Groves, farms, towns
Preferred nest site: Cavity in dead tree, utility pole
Clutch size: 4 to 7, usually 5

Incubation period: 14 days
Nestling period: About 27 days
Broods per season: 1 or 2
Preferred food: Insects, acorns, wild fruits

Lewis's Woodpecker

LEWIS'S woodpecker, a mountain species, is usually encountered from the Rockies westward, though it sometimes does visit the Plains. It also may be discovered in towns and suburbs. It has a distinct fondness for fruits in addition to acorns and insects, and is one of the mul-titude of birds attracted by cherries. Its nickname, "crow woodpecker," refers to its habit of flight, not to its size. It's just a bit larger than the hairy woodpecker.

Size: 9"L
Breeding range: Central British Columbia through southwestern Alberta and Montana to the Black Hills, south to northern New Mexico and Arizona to central California

Winter range: Southwestern Oregon to southwest Colorado, south to extreme northwest Mexico
Preferred habitat: Open woods to suburbs and towns
Preferred nest site: Cavity in tree

Clutch size: 6 to 8
Incubation period: 14 days
Nestling period: 21 days
Broods per season: 1
Preferred food: Sap, insects

Gila Woodpecker

Size: 8¼"L
Breeding range: Southeastern California, central Arizona, extreme southwestern New Mexico to Mexico
Winter range: Permanent resident in breeding range
Preferred habitat: Deserts, river groves, towns

Preferred nest site: Cavity in saguaro or cottonwood
Clutch size: 3 to 5
Incubation period: 14 days
Nestling period: Not known
Broods per season: 2, perhaps 3
Preferred food: Flying insects, ants, berries, corn

Picidae

Yellow-Bellied Sapsucker *(Sphyrapicus varius)*

THE yellow-bellied sapsucker, between sparrows and robins in size, is found over most of the country. Besides sap and insects, it eats fruits and berries. At the feeder it is more likely to accept apple bits, cherries, or strawberries than anything else except an occasional sample of suet or nutmeats.

The tongue of the sapsucker is brushlike and therefore not adapted to gathering the kinds of insects that many woodpeckers eat. It's perfect for its purpose, though. The bird sometimes gets quite boisterous from imbibing fermented sap, especially in spring. Margaret Millar, in *The Birds and the Beasts Were There*, tells a wonderful story of an inebriated sapsucker vigorously defending his source against a hopeful oriole.

Size: 9"L
Breeding range: Newfoundland west to Yukon; south through eastern British Columbia to Nevada, central Arizona, and southern New Mexico; east to mountains of Georgia, Virginia, and Massachusetts

Winter range: New Jersey west to southern California, south to Central America
Preferred habitat: Woodlands, groves, orchards
Preferred nest site: Cavity in tree

Clutch size: 4 to 7, usually 5 or 6
Incubation period: 12 to 14 days
Nestling period: 24 to 26 days
Broods per season: 1
Preferred food: Sap, insects, fruits, berries

Downy Woodpecker   (*Picoides pubescens*)

PROBABLY the most familiar of the woodpeckers is the downy woodpecker, a favorite at most feeders. This sparrow-sized black and white bird is found all over the United States except in southern Texas and the extreme Southwest. Its preferred feeder food is suet. Friendly and tame, the downy is found often near man, and city parks, open woods, farmyards, or door-yards are all equally appealing to it. About three-fourths of its natural food is insects such as beetles, caterpillars, wood-boring ants, larvae, and eggs. It rounds out its diet with seeds, berries and fruit, some sap, and the cambium layer of bark. In winter, downy woodpeckers are frequently found in loose flocks with nuthatches and chickadees.

The downy woodpecker defends an area of about four acres, about eight times that of his larger cousin, the hairy woodpecker. In courtship he merely flaps from tree to tree and sails in a deep loop. Once mated, the parents make a gourd-shaped nest, 8 to 12 inches deep, with a 1½-inch entrance hole, 5 to 30 feet from the ground. They brood their young for fourteen or fifteen days.

Size: 5¾"L
Breeding range: Newfoundland west to central Alaska, south to California, northern Arizona and New Mexico, through east Texas to the Gulf and to Florida
Winter range: Permanent

residents in breeding range
Preferred habitat: Mixed woodlands to suburbs and towns
Preferred nest site: Cavity in tree; accepts birdhouse
Clutch size: 3 to 6, usually 4 or 5
Incubation period: 12 days

Nestling period: 20 to 22 days
Broods per season: 1, 2 in the south
Preferred food: Insects, sunflower seeds

Hairy Woodpecker

(Picoides villosus)

HAIRY woodpeckers closely resemble the downies, but the hairy woodpecker is robin-sized and has a much stouter, heavier bill than the downy. It's also considerably less common. Though it's found in the same range — in fact, all over the country, including southern Texas and the Southwest — it's less likely to make its home close to ours. During breeding season, hairy woodpeckers prefer deep woods. The hairy woodpecker defends an area of a quarter acre around the nest site by drumming, chasing, and perching, but it explores throughout a larger area. When courting, male hairy woodpeckers drum, "still-pose" (just what it sounds like), and make fluttering flights.

Ornithologists think the female hairy woodpecker selects the site for a nest, usually new

for each season, but both parents excavate. It takes them one to three weeks to fashion the gourd-shaped nest, which is 5 to 60 feet from the ground and a foot deep, with an entrance hole 1⅞ by 1½ inches.

The female lays her eggs in wood chips.

The parents rear only one brood a year, and the male frequently makes himself a bedroom in a tree adjacent to the nesting site. The nestlings stay in the nest for three or four weeks (longer than downies) and, like other woodpecker young, are remarkably unlovely. But then, many birds are homely when hatched, and most of them go through one stage or another of looking at least awkward, gawky, or disheveled.

Hairy woodpeckers are most likely to come to feeders during the winter and will appreciate meat bones in addition to suet. Their major food is wood-boring beetles, but they also like beechnuts, hazelnuts, acorns, pine seeds, and piñons. They are less often seen with chickadees and other smaller birds than are the downies; but then, they're less often seen at all.

Size: 7½"L
Breeding range: Newfoundland west to Alaska, south to California and east to Florida, except arid portions of southwest U.S.
Winter range: Permanent resident in breeding range

Preferred habitat: Open woodlands with mature living and dead trees, wooded swamps, residential areas
Preferred nest site: Cavity, preferably in living tree
Clutch size: 3 to 6, usually 4

Incubation period: 11 to 12 days, 15 days also reported
Nestling period: 28 to 30 days
Broods per season: 1, 2 in the South
Preferred food: Adult and larval beetles, ants, fruits, nuts, corn, sunflower seeds

TYRANT FLYCATCHERS

THESE birds are migrants, who return to their breeding range in the spring when the insect population emerges. They are extremely important as devourers of insects, which make up ninety percent of their diet. *See* chapter 5 for tips on housing phoebes.

Left to right: Eastern phoebe, eastern kingbird.

TYRANT FLYCATCHERS   *Tyrannidae*

Eastern Phoebe   *(Sayornis phoebe)*

EASTERN phoebes prefer shelflike projections on bridges or buildings as nest sites. The female builds a circular or semicircular nest in anywhere from three to thirteen days. It is large — 4½ inches in diameter and 4 inches high — and well made of weeds, grasses, and mud covered with moss and lined with fine grasses and hair.

Size: 7"L
Breeding range: Nova Scotia west to extreme southeastern Yukon; through central Canada south, east of the Rockies, to northeastern New Mexico; though central Texas to mountains of central Georgia

Winter range: Maryland south along Atlantic coast to Florida, west along coastal Gulf states to southwestern New Mexico and south to Mexico
Preferred habitat: Open woodlands, farms, towns, roadsides

Preferred nest site: Ledge, bridge, building
Clutch size: 3 to 8, usually 5
Incubation period: 15 to 17 days
Nestling period: 15 to 17 days
Broods per season: 2
Preferred food: Flying insects

Eastern Kingbird

(Tyrannus tyrannus)

THE eastern kingbird is well known for being an aggressive bird. In defense of his half acre, he will attack crows, hawks, and owls even a hundred feet over his territory. He also proclaims his property by what is called the *tumble-flight,* in which he flies very high, glides down in stages, and then tumbles through the air.

Even in courtship the male seems hostile at first, as if he were just as willing to drive the female away from his territory as entice her to share it. The wing flutter is characteristic of his courtship.

The pair builds their rough, bulky nest on a tree limb away from the trunk, using weeds and grasses and lining it with fine grasses. The dimensions are 5½ inches in diameter and 3¼ inches high.

The nestling period of the eastern kingbird is fourteen to seventeen days. For the first few days the female alone feeds the chicks; then both parents do. They carry away the fecal sacs and drop them beneath a perch. The fledgling period lasts for two to three weeks, and the family stays together until migration.

Size: 6¾"L
Breeding range: Nova Scotia west to southeastern Yukon and British Columbia; south through Washington, eastern Oregon, and northern Nevada to Texas panhandle; and south to the Gulf states and Florida

Winter range: Central and South America
Preferred habitat: Wood edges, farms, roadsides
Preferred nest site: Tree, bush, post
Clutch size: 3 to 6, usually 3

Incubation period: 12 to 13 days
Nestling period: 13 to 17 days
Broods per season: 1
Preferred food: Flying insects, wild fruits

SWALLOWS

EXPERT flyers and voracious insect eaters — like hummingbirds and swifts, the members of the family Hirundinidae prefer life in the sky. They fly within three weeks of hatching and are skilled flyers as soon as they leave the nest. The parents force insects down their throats until they can catch their own on the wing. A single nest of swallows may eat several thousand insects daily.

Swallows may use the previous year's nest — a deep cup lined with grasses and feathers.

Left to right: Purple martin pair, tree swallow, barn swallow pair.

SWALLOWS

Hirundinidae

Barn Swallow

(Hirundo rustica)

BARN swallows often nest in colonies around or in buildings, bridges, and culverts. (Formerly they used caves in cliffs.) They plaster the mud-and-straw structures to beams and use feathers, preferably white, for the lining. The top of the nest is semicircular, 5 inches in diameter, and it tapers downward in a cone shape. It takes them six to eight days to build a new one, less time to repair an old one. If bad weather comes (that is, too cold for insects), barn swallows simply leave the nest and their nestlings. When they come back, they pitch out the corpses and start all over again.

Size: 6–7"L
Breeding range: Newfoundland west to southeastern Alaska, south to Mexico and east to Alabama and the Atlantic, locally in the Gulf states
Winter range: South America

Preferred habitat: Farmlands and residential areas
Preferred nest site: On, in, or around buildings, often in colonies
Clutch size: 4 to 6, usually 4 or 5

Incubation period: 15 days
Nestling period: 16 to 23 days
Broods per season: 1 or 2
Preferred food: Insects

Tree Swallow

(Tachycineta bicolor)

TREE swallows may move into a nesting box meant for bluebirds. If so, don't be disappointed. They are delightful tenants with their swooping flight and appetite for mosquitoes, and they will often let you approach to see their brilliant blue-black plumage and cream breasts up close.

When courting, tree swallows engage in flutter flight, bowing, and billing. Once mated, they use a tree hole or nestbox 5 to 10 feet from the ground with a 1½-inch entrance hole. It takes the female up to two weeks (some sources say up to a month) to build the nest of dry grasses lined with feathers; the male gathers some of the material. Tree swallows may be solitary nesters or they may nest in groups. They defend the immediate area of their nests by chasing intruders.

The parents brood their nestlings for the first three days of the three-week nestling period. Both parents feed, and both remove fecal sacs, which they drop in water if possible. The fledgling period lasts two or three weeks.

See Chapter 5 for more information on providing nestboxes.

Size: 5"L
Breeding range: Labrador to northern Alaska and south to California, northern New Mexico, Mississippi Valley to central Mississippi northeast to Virginia
Winter range: Coastal areas from Virginia south, along Gulf states to central California, south to Mexico and Central America
Preferred habitat: Open, wooded swamps and near water
Preferred nest site: Natural cavity in tree, old woodpecker hole, birdhouse

Clutch Size: 4 to 7, usually 5 or 6
Incubation period: 13 to 16 days
Nestling period: 16 to 24 days
Broods per season: 1
Preferred food: Flying insects, berries, seeds

Purple Martin

(Progne subis)

PURPLE martins nest in dense colonies, 15 to 20 feet from the ground. The male arrives in late winter and finds a martin house (*see* Chapter 5); the female arrives and selects a room. Both parents build the nest, using grasses, twigs, bark, paper, leaves, and string. Both parents incubate the eggs and care for the chicks, catching huge quantities of flying insects to feed the nestlings. In late summer martins gather in enormous flocks to fly south.

Size: 7"L
Breeding range: Newfoundland west to southern Ontario, west to British Columbia, south to central Texas and east to Florida. Also southwest British Columbia to Mexico
Winter range: Brazil

Preferred habitat: Farmlands, parks, residential areas near water
Preferred nest site: Previously in tree cavities, now almost exclusively in birdhouses; colonial
Clutch size: 3 to 8, usually 4 or 5

Incubation period: 15 to 18 days
Nestling period: 26 to 31 days
Broods per season: 1, usually
Preferred food: Flying insects

Cliff Swallow

(Hirundo pyrrhonota)

Size: 5"L
Breeding range: Newfoundland west to northern Alaska, south to Central America, east to Missouri and Virginia (absent from Gulf states)
Winter range: South America

Preferred habitat: Open to semiwooded land, farms, rivers and lakes, villages
Preferred nest site: Under bridge, eaves of buildings, cliff; almost always colonial

Clutch size: 3 to 6, usually 4 or 5
Incubation period: 15 to 16 days
Nestling period: 24 days
Broods per season: 1 or 2
Preferred food: Flying insects

THE NOISY SET

THE family Corvidae, of which crows, jays, and magpies are members, is not likely to be overlooked. Some of these birds are strikingly handsome; all are conspicuous.

Most people never have a chance to hear any of the soft, melodic songs of **jays** but are quick to recognize their less mellifluous shrieks and jeers. Their cries frequently alert other wildlife to the presence of danger.

Jays are, for all practical purposes, omnivorous and come readily to feeders, where they tend to drive other birds away. Since they're quick to discover new food sources, they frequently are among the early arrivals at a freshly installed feeder.

Occasionally **crows** come to feeders, but few of us encourage them. For one thing, a crow needs about a pound and a half of food daily. They prefer to eat on the ground, which is just as well, considering their size. While there is no reason to persecute these birds, neither is there reason to encourage them. They're capable of fending for themselves under most conditions.

Clockwise from lower left: Common raven, gray jay, blue jay, black-billed magpie, Steller's jay, American crow, scrub jay, Clark's nutcracker (center).

186

Gray Jay (*Perisoreus canadensis*)

GRAY jays, also called Oregon jays and Canada jays, are common in northern woods and mountainous areas. They have no crest and resemble large, disheveled chickadees. They're also called "whiskey jacks," "meat birds," and several other names, some of them unprintable. Inveterate thieves, they steal food from camps and bait or game from traps. Like other jays, they often take away more food than they can eat and store it. They're especially partial to baked beans and meat, raw or cooked. They're as willing to take these goodies from a camper's table as from a feeding tray.

Gray jays nest early — late February to April — and build bulky, high-walled nests of twigs, grasses, and bark lined with moss, lichens, plant down, and feathers. Nests are held together with cocoons and spider webs and are usually 4 to 10 feet from the ground, though they have been found as high as 30 feet up.

Size: 10"L
Breeding range: Labrador west to northern Alaska, south to northern California, Rockies, and New Mexico; Black Hills, northern Michigan and New York; northern New England
Winter range: Permanent resident in breeding range; some wandering south, to Pennsylvania and southern Great Plains
Preferred habitat: Coniferous forests
Preferred nest site: Conifer
Clutch size: 2 to 5, usually 3 or 4

Incubation period: 16 to 18 days
Nestling period: 15 days
Broods per season: 1
Preferred food: Omnivorous — insects, fruits, seeds, buds

Scrub Jay

(Aphelocoma coerulescens)

THE scrub jay lives in the chaparral country of the West. A subspecies is the Florida jay, found only in Florida. It is a crestless bird, blue and gray with no white markings. It's not quite as strident in voice as other jays. Scrub jays feed on acorns, peas, corn, grain, and nuts as well as ants, wasps, and other animal food. They'll come to feeders and especially like peanuts and suet. They tame easily.

The territory of the scrub jay consists of a single scrub oak. Scrub jays may nest alone or in scattered colonies with up to six nests. Together a pair forms oak twigs into a thick-walled cup lined with rootlets. It takes them about five days, and they prefer a lower site — 4 to 12 feet high — than blue jays. In Florida, the nest is small and well built, and the walls may contain stems and leaves as well as twigs; the lin-

ing may have wool, moss, and feathers in addition to root-lets. In the West, scrub jays build a bulkier nest and add horsehair to the lining.

Florida scrub jays nest one pair to a scrub oak, and the youngsters have to wait until an adult dies to claim an oak of their own. Until that time, they help their parents with whatever brood they're raising and do not breed themselves.

Size: 10"L
Breeding range: Southwestern Washington south to southern Mexico; central Florida
Winter range: Permanent resident in breeding range

Preferred habitat: Brush, river woods, junipers
Preferred nest site: Bush or low tree
Clutch size: 2 to 6
Incubation period: 16 to 19 days

Nestling period: 18 days
Broods per season: 1 or 2
Preferred food: Omnivorous — insects, acorns, young birds

Corvidae

Blue Jay

(Cyanocitta cristata)

OUR two crested jays are the blue jay and Steller's jay. Generally speaking, their ranges don't overlap. The blue-and-white blue jay occurs east of the Rocky Mountains; the black-and-blue Steller's jay, from the Rockies westward.

Blue jays are abundant in their range, in yards with or without feeders, in parks and any other place they may find scraps left by picnickers. They also are road-kill scavengers. They tend to feed earlier in the day than many of the other feeder birds, and they prefer ground feeding. Swinging feeders discourage them. Their food preferences are corn, peanuts, nutmeats, suet, and sunflower seeds. They aren't fond of the smaller seeds. Plant food supplies 90 percent of their needs in winter and 60 percent in summer. To the fruits, berries, and nut in season they'll add ants, snails, and small fish and frogs.

The courtship of blue jays is surprisingly inconspicuous and may consist primarily of bobbing and mate feeding. It isn't unusual to see a flock of males chasing a single female at breeding time. (Such chasing is noisy, but what else would you expect from jays?) Their numbers gradually diminish until a single pair remains — silent. Such chases may involve first-time breeders.

Blue jays build coarse, careless, cup-shaped nests with ragged brims hidden in a crotch or outer branch, usually 10 to 20 feet from the ground, although the height may vary from 5 to 50 feet. They prefer evergreens. Both parents build, with thorny twigs, rootlets, bark, moss, leaves, paper, rags, and string and a lining of bark, grasses, leaves, and feathers. Nests are 7 to 8 inches in diameter, 4 to 4½ inches high. Both male and female gather materials from the ground, but the males also break twigs from trees. The building takes about five days, and the female does a bigger share of work than the male.

Female blue jays brood their young for the first few days. Both parents feed the young;

Size: 10"L
Breeding range: Newfoundland west to southern Alberta, south to Texas and Florida
Winter range: Permanent resident in breeding range, except northernmost portions

Preferred habitat: Woodlands, parks, gardens, cities
Preferred nest site: Tree
Clutch size: 3 to 7, usually 4 or 5
Incubation period: 17 to 18 days

Nestling period: 17 to 21 days
Broods per season: 1, 2 in the South
Preferred food: Omnivorous — seeds, fruits, acorns, young mice, nestlings

the male also feeds the female. Although blue jays have ill-defined territories, they are aggressive toward interlopers, especially squirrels and cats. Their usual attack is dive-bombing.

I miss the blue jays while they're busy during the summer raising their families (and, some say, preying on other nests). Then along about August we start hearing their cries, and soon the impudent, good-looking rascals are at the feeders again, commenting derisively on our offerings. It's probably a character fault that makes me enjoy their foibles so thoroughly.

In the East, blue jays have served as experimental subjects. (It's significant, I think of this family's reputation for intelligence that they're used in this way.) They have demonstrated, to the satisfaction of one research study at least, an ability to concentrate. No one who has watched them around the garden is apt to deny it.

JAYS, MAGPIES, AND CROWS   *Corvidae*

Steller's Jay   *(Cyanocitta stelleri)*

STELLER'S jays like nuts, acorns, and fruit. Sometimes their appetite for fruit gets them into trouble with orchardists. Like blue jays, they have the reputation of occasionally eating the eggs and nestlings of smaller birds. In an effort to dissuade them, offer them eggshells at the feeding station.

Steller's jays build a large, bulky, bowl-shaped nest of sticks, mud, and grasses lined with rootlets, pine needles, and grasses on a platform of old leaves in the crotch of a tree from 2 to 100 feet from the ground, but most often 8 to 25 feet up. Steller's jays remain in the nest for eighteen to twenty-one days.

Size: 11"L
Breeding range: Southern Alaska east to southwestern Alberta, south through the mountainous regions of western United States, south to Nicaragua

Winter range: Permanent resident in breeding range
Preferred habitat: Coniferous and pine/oak forests
Preferred nest site: Conifer
Clutch size: 3 to 5, usually 3 or 4

Incubation period: 16 to 18 days
Nestling period: 18 to 21 days
Broods per season: 1
Preferred food: Omnivorous — acorns, fruits, seeds, berries

190

Black-Billed Magpie

(Pica pica)

MAGPIES, just about crow-sized, are striking, long-tailed, black and white birds seldom seen at feeders, although they are frequently found around western farms and ranches. Their nests have been described as resembling bushel baskets in size and shape, made of twigs and sticks and lined with finer materials and mud.

Like most of this voracious family, magpies eat a wide variety of foods, their prey ranging from reptiles, crustaceans, and scorpions to rodents and a wide variety of insects. Magpies are not popular with farmers, however, who accuse them of marauding baby chicks and bothering the young of farm animals.

Size: 18"L
Breeding range: South and central Alaska to western Ontario, south to northern New Mexico, northwest to Oregon, east central California
Winter range: Permanent resident in breeding range

Preferred habitat: Roadsides, thickets, fields, ranches
Preferred nest site: Bush or tree, often in scattered colony
Clutch size: 6 to 13, usually 7
Incubation period: 16 to 18 days

Nestling period: 22 or more days
Broods per season: Probably 1
Preferred food: Omnivorous — insects, vegetable matter, carrion

191

American Crow

(Corvus brachyrhynchos)

CROWS are perhaps the best-known bird on the continent. Often maligned, they're intelligent and can be easily tamed. Despite campaigns that have been waged against them, they're probably more numerous now than when the colonists arrived.

Courtship among crows begins with fighting in the flock. Considering their usual behavior, that isn't surprising. But can you believe that mated crows bill like doves? During courtship they perch on a limb together. The male walks toward the female, bows, ruffles his feathers, and spreads his wings and tail. He lifts his head up and then lowers it. It's all very courtly, and his lady is impressed.

Crows build large, bulky nests. Cup-shaped and made of sticks, they are lined with bark, vegetable fibers, rootlets, grasses, and moss. At times they use feathers, corn-stalks, string, cloth, and seaweed. They are usually 30 feet from the ground, but the height may vary from 10 to 100 feet. They particularly like pines or oaks near a clearing. The nests are 7 inches high and about 2 feet in diameter. Sometimes a crow is evicted by a great horned owl; other times a long-eared owl may use an abandoned crow's nest. Crows defend only the immediate area of the nest; these social creatures flock together in huge communal roosts.

Because crows begin incubation as soon as the first egg is laid, hatching occurs over a period of days. Parents share both incubation and care of the young. The female broods the young for the first ten days of the five-week nestling period; both parents feed them. During the fledgling period of about two weeks, the young follow the parents and beg.

Size: 17"L

Breeding range: Newfoundland west to British Columbia, south to northern Baja California, northeast through central Arizona and northwestern New Mexico to Oklahoma, and south through east Texas to Gulf states and Florida

Winter range: Mainly south of Canada

Preferred habitat: Woodlands, farmlands, shores

Preferred nest site: Crotch of tree

Clutch size: 3 to 9, usually 4 or 5

Incubation period: 18 to 21 days

Nestling period: 25 days

Broods per season: 1, 2 in the South

Preferred food: Omnivorous — grains, insects, carrion

Clark's Nutcracker *(Nucifraga columbiana)*

CLARK'S nutcrackers are well-known, jay-size gray-and-black birds of the western mountains. Practically a tourist attraction in the Rockies, they're fond of meat and peanuts, which will lure them to feeders in their range. They have acquired a reputation as camp robbers. Like the gray jay, or "Whiskey Jack," they're both bold and inquisitive — and quite willing to engage in thievery.

Clark's nutcracker has been the subject of research on animal memory. Like other Corvidae family members, it stores food, and scientists have discovered that the birds remember where they put it. Apparently they make clusters of caches and utilize large landmarks as reminders. They clean out one batch before starting on another.

Not much is known of the courtship of Clark's nutcracker, which is assumed to be even quieter than that of the jays. Their nests are found under an evergreen branch and are similar to those of the jays.

Size: 11"L
Breeding range: High mountains from central British Columbia to Baja California, east to northwest New Mexico, north to the Black Hills; northern Nuevo Léon
Winter range: Permanent resident in breeding range
Preferred habitat: High mountains, coniferous forests
Preferred nest site: Conifer
Clutch size: 2 to 6, usually 2 to 4
Incubation period: 17 to 22 days
Nestling period: 18 to 28 days
Broods per season: 1, sometimes 2
Preferred food: Omnivorous — piñon nuts

Common Raven

(Corvus corax)

RAVENS have retreated northward during this century, partly because humans have destroyed them through fear that they would attack young farm animals. Yet they are useful predators of rodents and exceptionally clever birds.

Solitary nesters, ravens build their large structures in coniferous trees or on cliffs or rock ledges, often on top of last year's nest. Their deeply hollowed nests are built of branches, sticks, twigs, and grapevines and thickly lined with hair, moss, grasses, and bark shreds. They are 6 inches deep and 2 to 4 feet in diameter.

Size: 26½"L
Breeding range: Northwestern Alaska to Greenland, south to northern border of United States; found also in southern Appalachians
Winter range: Permanent resident of breeding range

Preferred habitat: Forest; seacoast; tundra
Preferred nest site: Tall tree or rocky cliff
Clutch size: 5 to 7
Incubation period: 20 to 21 days
Broods per season: 1

Preferred food: Mice, rats, insects

CHICKADEES AND TITMICE

THESE charming birds are part of the Paridae family, one dear to the hearts of bird lovers. They come to windowsills with complete confidence and can even be persuaded to eat from the hand. This will, however, take patience.

You'll hear the Paridae family referred to both as the titmouse family and the chickadee family. It also contains verdins. All of these birds are prodigious consumers of small insects and their eggs, winter and summer. They eat some seeds and berries, too, but in relatively small numbers.

Both chickadees and titmice rely on insect food as their staple diet item, eating caterpillars, wasps, beetles, and various other insects and their eggs. When planning your garden, you might want to keep in mind that the vegetable food they prefer includes almonds, acorns, cashews, beechnuts, butternuts, mulberries, honeysuckle, wax myrtle berries, and the seeds of locusts.

Left to right: Tufted titmouse, Carolina chickadee, mountain chickadee, black-capped chickadee, boreal or brown-capped chickadee.

Paridae

Black-capped Chickadee   (*Parus atricapillus*)

BLACK-capped chickadees are found all over the northern part of the continent from coast to coast and are year-round residents throughout most of their range. They're woodland birds that will often move into orchards or shade trees. Their natural food is largely insects, even in winter (when they feed on eggs and pupae), but they do eat some seeds, fruits, and berries.

To attract black-capped chickadees to your garden, you need trees. They like balsam firs, locust, birch seeds

and buds, butternuts, blueberries, bayberries, and the berries of poison ivy. Their inquisitive and trusting nature often leads them to be among the first visitors to a new feeder. It should come as no surprise, considering the proportion of animal food in their natural diet, that they like suet, but they are very fond of sunflower seeds as well as squash and pumpkin seeds, peanuts or peanut butter, butternuts, and oats. They sometimes develop a liking for honey and syrup.

From August to February, black-capped and Carolina chickadees have a feeding territory of about twelve to twenty acres, which is defined by the flock using it. Except in

the northern sections of their range, where chickadees tend to be migratory, these small resident flocks consist of the same birds throughout the year. Within the flock there is a dominance hierarchy. A pair found around a feeding station in summer is very likely the dominant pair of the winter flock. Birds feeding together may be "friends" or mates.

Chickadees breed, however, in isolated pairs. From March to July, the *fee-bee* song of the male announces a breeding territory of a half acre to about ten acres, which he defends aggressively with short chases of rivals and occasional fights. Wing fluttering is the principal courtship behavior of black-capped chickadees, occurring from March through June, after the breakup of the winter flocks. The pair bill-touches and feeds each other.

Black-capped chickadees share in the excavation of a cavity (males and females carry the wood chips away

Size: 4½"L
Breeding range: Newfoundland west to central Alaska, south to northern California, southeast to northern New Mexico, northeast to New Jersey; mountains south to Georgia

Winter range: Permanent resident in breeding range
Preferred habitat: Mixed and deciduous woods, wood edges, gardens, towns
Preferred nest site: Cavity in dead tree or stump, birdhouse

Clutch size: 4 to 13, usually 6 to 8
Incubation period: 11 to 13 days
Nestling period: 16 days
Broods per season: 1 or 2
Preferred food: Insects, seeds, fruits

from the site and drop them), but the female makes the nest itself — of moss, vegetable fibers, cinnamon fern, wood chips, feathers, cocoons, hair, fur, and sheep's wool. They like a 1-inch entrance hole, in a site 4 to 15 feet from the ground. Sometimes, for reasons unknown, they abandon the nest and start a new one somewhere else. They are especially fond of birch snags, which are easily pierced and excavated although the papery bark holds the rotten wood together. They accept nesting boxes.

Black-capped chickadee parents feed their nestlings for a period of sixteen or seventeen days. The female broods for the first few days, during which time the male feeds her. Fledglings follow their parents for a few weeks, but after about ten days the parents stop feeding them. Eventually the parents drive the young away, and the juveniles find themselves a tiny territory and defend it — by singing quite inaccurately.

TITMICE AND CHICKADEES   *Paridae*

Mountain Chickadee   *(Parus gambeli)*

SIMILAR to these species is the mountain chickadee. A bit larger than the black-capped, it is most apt to be found in coniferous mountain areas from the Rockies west, except in the Pacific Northwest. Like Carolina and black-capped chickadees, its food is largely insects, but it also eats berries and seeds. At feeders, mountain chickadees particularly relish piñon nuts and suet. As friendly and curious as the others, mountain chickadees can become so tame that they will eat from your hand.

Mountain chickadees nest in evergreen forests in cavities 1 to 15 feet from the ground. The entrance hole is 1 inch in diameter, the cavity 9 inches deep. The nest is made of grasses, moss, plant down, bark, rootlets, squirrel and rabbit fur, cow or deer hair, and sheep's wool. It's thought that mountain chickadees feed their young by regurgitation for the first few days.

Size: 4¼"L
Breeding range: Southeastern Alaska to southwestern Alberta, south to west Texas and west through southeastern Arizona to northern Baja California; absent from coastal areas

Winter range: Permanent resident in breeding range
Preferred habitat: Conifers
Preferred nest site: Cavity in tree or stump, birdhouse
Clutch size: 5 to 12, usually 7 to 9
Incubation period: 14 days

Nestling period: To 21 days
Broods per season: 1, often 2
Preferred food: Insects, seeds, berries

197

Carolina Chickadee *(Parus carolinensis)*

CAROLINA chickadees are almost identical in appearance to black-capped chickadees but noticeably smaller. Their ranges overlap slightly, and the two may hybridize in the small range common to them. As its name implies, the Carolina chickadee is native to the Southeast.

The berries of poison ivy seem to be Carolina chickadees' favorite vegetable food, but they also like cashews, almonds, sweet gum seeds, acorns, and other seeds. Their chief animal food is moths and caterpillars, but their diet also includes plant lice, spiders, and leaf hoppers. At the feeder, try American and cottage cheese in addition to suet, peanut butter, nutmeats, and sunflower, squash, and pumpkin seeds.

Carolina chickadees like a cavity 6 to 15 feet from the ground. The nest is built on a foundation of moss, grasses, and bark and lined with vegetable down, feathers, hair, and fur. One side, higher than the others, serves as a flap that is pulled down over the eggs when a parent is off the nest.

The nesting behavior of the Carolina chickadee is identical to black-capped chickadees'. Parents feed their nestlings for a period of sixteen or seventeen days; the male feeds the female while she is brooding during the first few days; parents feed the fledglings for about ten days; and they eventually drive them off to find a territory of their own.

Size: 4¼"L
Breeding range: Southern edge of black-capped chickadee range, New Jersey west to Oklahoma, south to the Gulf states and east to Florida
Winter range: Permanent resident in breeding range

Preferred habitat: Swamps, woods, residential areas
Preferred nest site: Cavity in tree, birdhouse
Clutch size: 5 to 8, usually 6
Incubation period: 11 to 13 days
Nestling period: 17 days

Broods per season: 1
Preferred food: Insects, seeds, fruits

Boreal or Brown-Capped Chickadee   *(Parus hudsonicus)*

FEW of us will have the opportunity to feed the boreal, or brown-capped, chickadee, which lives in the great north woods. In extremely cold winters, our most northerly regions may have boreal chickadee visitors. Although it's most of all an insect eater — especially of aphids, beetles, ants, and wasps — the boreal chickadee consumes pine, balsam fir, and birch seeds, and cedar berries. At feeders it likes bacon grease, peanut butter, suet, and sunflower seeds.

Rather spectacularly, boreal chickadees mate-feed in mid-air during courtship, the female fluttering her wings and ruffling her feathers. They nest in cavities or stumps 6 inches deep and 1 to 10 feet from the ground and make their nests of moss, lichens, bark, and fern down, lining them with feathers or fur.

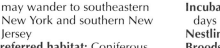

Size: 4¼"L
Breeding range: Labrador west to northwestern Alaska, south to western British Columbia, east to northern New York and Maine
Winter range: Permanent resident in breeding range; may wander to southeastern New York and southern New Jersey
Preferred habitat: Coniferous forest
Preferred nest site: Cavity in rotting tree or stump
Clutch size: 4 to 9, usually 6 or 7

Incubation period: 12 to 14 days
Nestling period: 18 days
Broods per season: 1
Preferred food: Insects, seeds, berries

Tufted Titmouse

THE best known of the titmice is probably the tufted titmouse, resident over much of the United States east from the Great Plains. It's our only small gray bird with a crest. Although the tufted titmouse was formerly absent north of New Jersey, its range seems to be extending, especially in New York State and southern New England. Particularly in winter, it's common to see titmice in the company of other woodland birds such as woodpeckers, nuthatches, chickadees, and creepers.

At feeders, titmice eat cantaloupe seeds, nutmeats, peanuts, peanut butter, suet, and safflower and sunflower seeds. Like the other acrobatic members of the family, they're happy to use a swinging feeder.

Courtship behavior of the tufted titmouse is unknown, which we can count as one blow for privacy. Some believe it is ritual feeding of the female and wing fluttering of both male and female. I suspect such conjecture is based on the known behavior of some of their relatives, such as the chickadees.

Tufted titmice are cavity nesters who like a 1½-inch entrance hole 3 to 30 feet from the ground, though they're known to nest as high as 97 feet up. Both male and female build the cupped nest, made of leaves, bark, moss, grasses, cotton, wool, string, cloth, feathers, fur, and old snakeskins. (They even pluck hair from live animals for their nests.) They will accept nestboxes.

During the fifteen to sixteen days of the nestling period, female tufted titmice brood their young, and the male feeds the female while she's on the nest. Both parents feed the young.

Size: 5½"L
Breeding range: Central New England west to northern Nebraska, south to Texas, the Gulf states, and Florida
Winter range: Permanent resident in breeding range

Preferred habitat: Swamps, deciduous woodlands, shade trees
Preferred nest site: Natural cavity in tree, birdhouse
Clutch size: 4 to 8, usually 5 or 6

Incubation period: 12 to 14 days
Nestling period: 15 to 18 days
Broods per season: 1
Preferred food: Insects, seeds, mast, fruits

Bushtit

(Psaltriparus minimus)

THESE two species are not likely to visit feeders, but they might visit gardens or parks.

The **bushtit** might build a nest in a brush pile; the **plain titmouse** is a cavity nester and might accept a nesting box.

Size: 3½"L
Breeding range: Southwestern British Columbia along coast to southern Baja California, south central Washington through southern Idaho to eastern Colorado, west Texas

to northern Mexico
Winter range: Permanent resident in breeding range; wanders in winter
Preferred habitat: Scrub, mixed woods, parks, city gardens
Preferred nest site: Bush, tree

Clutch size: 5 to 15, usually 5 to 7
Incubation period: 12 days
Nestling period: 14 to 15 days
Broods per season: 2
Preferred food: Insects and larvae

Plain Titmouse

(Parus inornatus)

Size: 5"L
Breeding range: Southern Oregon to south central Colorado, south to southern New Mexico and west Texas, west to northern Baja California, southern Baja California

Winter range: Permanent resident in breeding range
Preferred habitat: Woods, shade trees
Preferred nest site: Cavity in tree, birdhouse

Clutch size: 6 to 12, usually 6 to 9
Incubation period: 14 days
Nestling period: 16 days
Broods per season: Probably 1
Preferred food: Insects, acorns, berries

NUTHATCHES

THAT avian acrobat, the nuthatch, is one of our common feeder birds. Its family, the Sittidae, has thin bills, sturdy bodies, short squared tails, and strong feet and legs. They're distinctive in appearance but even more so in behavior. They consistently go down tree trunks headfirst. All nuthatches will use swinging feeders and have no hesitation about coming to windowpane feeders.

Nuthatches are fascinating to watch and have been given numerous nicknames, including "upside-down bird" and "twirl-around-a-twig."

Especially useful in orchards, nuthatches eat many kinds of injurious insects, although beneficial insects appeal to them as well. They are cavity nesters, and the female does most of the incubating.

Clockwise from left: Red-breasted nuthatch, brown creeper, white-breasted nuthatch, brown-headed nuthatch; center, pygmy nuthatch.

White-Breasted Nuthatch   *(Sitta carolinensis)*

THE white-breasted nuthatch is resident over most of the country throughout the year. About half of its diet is animal, consisting of ants, beetles, flies, locusts, spiders, and scale insects as well as insect eggs. The plant part includes beechnuts, acorns, and hickory nuts, pine, fir, and maple seeds, mountain ash and juniper berries, apples, and sunflower seeds. Nuthatches store food. They are very fond of suet but also visit feeders for the seeds of sunflower, squash, and pumpkin. They'll nibble on carrots if you'll supply them. White-breasted nuthatches are often found in the company of kinglets, downy woodpeckers, chickadees, and creepers.

Ours are so tame, they merely move to the other side of the feeder and peer at us from underneath if we go out to fill it while they're dining.

White-breasted nuthatches are examples of birds that mate for life rather than for the breeding season. The male in courtship bows, sings, and pretends to intimidate the female. During the breeding season, white-breasted nuthatches claim a

territory of twenty-five to fifty acres.

White-breasted nuthatches want a hole 2 to 60 feet from the ground and 6 inches deep and will accept nestboxes. The female builds the nest of loose bark with grasses, twigs, rootlets, bark, and feathers on top, then lined with hair sometimes plucked from a live animal. The male brings the female the materials and feeds her while she is on the nest.

Size: 5"L
Breeding range: New Brunswick west to southern British Columbia, south to mountains of Mexico
Winter range: Permanent resident in breeding range; wanders in winter

Preferred habitat: Mixed forests to suburbs
Preferred nest site: Cavity in tree, birdhouse
Clutch size: 4 to 10, usually 8
Incubation period: 12 to 13 days
Nestling period: 14 days

Broods per season: 1
Preferred food: Insects, seeds, fruits, mast

203

Red-Breasted Nuthatch

(Sitta canadensis)

THE red-breasted nuthatch is found in the same range as the white-breasted, but its numbers seem to fluctuate greatly. It has been postulated that this is a result of fluctuations in its favorite foods. The seeds of spruce, fir, and maple are preferred. In times of heavy snow cover especially, it feeds on the seeds of weeds such as ragweed and dock. It drinks sap from the holes left by sapsuckers. A tame creature, it comes to suet, sunflower seeds, and nutmeats at feeders. It can be trained to take them from the hand. Like the white-breasted nuthatch, it carries food away from feeders to hide it.

The male red-breasted nuthatch's courtship ritual is more complicated than that of the white-breasted. He struts before the female, flirting his wings and tail and bowing. As an added fillip, he flies in an ellipse 100 feet long above the tops of the pines.

Red-breasted nuthatches accept nestboxes. They need entrance holes an inch in diameter (around which they smear pitch, for some reason). The nest is of wood chips, sometimes with feathers, grasses, rootlets, and bark added.

Size: 4"L
Breeding range: Labrador west to southern Alaska, south to southern California, east to central New Mexico, north to western Montana, northern Minnesota east to New Jersey, mountains to Georgia

Winter range: Wanders irregularly to southern Arizona, New Mexico, and Texas, the Gulf states and Florida; retreats from more northerly part of range
Preferred habitat: Conifer forests

Preferred nest site: Cavity in dead conifer, birdhouse
Clutch size: 4 to 7, usually 5 or 8
Incubation period: 12 days
Nestling period: 21 days
Broods per season: 1
Preferred food: Insects, seeds

Brown-Headed Nuthatch *(Sitta pusilla)*

THE brown-headed nuthatch is found most often in pine woods in the southern states. Especially fond of pine seeds, it also eats nuts, ants, scale insects, and insect eggs. It comes to feeders for suet, nutmeats, and sunflower seeds.

Size: 3½"L
Breeding range: Coastal Delaware south to Florida, west to eastern Texas
Winter range: Permanent resident in breeding range

Preferred habitat: Pine woods, cypress swamps
Preferred nest site: Dead pine, post, utility pole
Clutch size: 5 or 6
Incubation period: 14 days

Nestling period: Not known
Broods per season: 1
Preferred food: Insects, pine seeds

Pygmy Nuthatch *(Sitta pygmaea)*

THE pygmy nuthatch is actually the same size as the brown-headed, and there is some suspicion that it's a subspecies. Found only from southwestern South Dakota westward, it, too, prefers coniferous trees as a habitat.

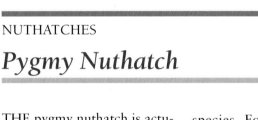

Size: 3½"L
Breeding range: Southern British Columbia east to western Montana, south to central Washington and south to Baja California, Black Hills southwest to eastern Nevada, south to Mexico
Winter range: Permanent resident in breeding range
Preferred habitat: Pine forests
Preferred nest site: Cavity in conifer

Clutch size: 4 to 9
Incubation period: About 14 days
Nestling period: Not known
Broods per year: 1
Preferred food: Insects

Brown Creeper

(Certhia americana)

THE family to which brown creepers belong, Certhiidae, occurs all over the northern part of the world. Creepers are smaller than a sparrow and are frequently overlooked because they blend so well with their usual background.

That usual background is a tree trunk, which the creeper ascends in spirals to a height of about twenty feet and then flutters down to the next tree and starts all over again. It's after tiny insects and their eggs and larvae. The creeper's slender, curved bill is well suited for the occupation and its stiff tail serves as a prop. They don't damage any crops, and their relentless search for insects yields dormant ones in winter as well as tiny ones that birds of the woodpecker tribe overlook.

Their activities are so re-

stricted — up and down, up and down, searching trees for food — that they occasionally behave peculiarly. If during their travels they reach a treeless area, they'll creep up any vertical object available, whether it be a fence post, telegraph pole, or brick wall.

Except during their migrations, you're unlikely to see many creepers together at any given time. They are most likely to be seen in winter except in the far north or in high altitudes. During the winter months they may feed with flocks of chickadees and nuthatches, on the fringes of the group. They're not particularly gregarious. They'll come to a feeder for suet and

chopped peanuts, and they like peanut butter, which some people spread on the trunks of trees for the convenience of the little bird.

Once it was believed that brown creepers were cavity nesters, but it's known now that they nest under loose flakes of bark on a tree, 5 to 15 feet from the ground. The male gathers materials for the nest, and the female builds, taking up to a month. The shape of the nest, made of a foundation of twigs, leaves, and bark lined with grasses and moss, conforms to the space. Its center is neatly cupped; the sides have points. Brown creepers are thought to be monogamous.

Size: 4¾"L
Breeding range: Newfoundland west to Alaska, south to mountains of southern California to Mexico, north to southern Alberta, east to New Jersey, mountains to Georgia
Winter range: Pennsylvania

northwest to southern Alberta, south to northern Mexico, east to Florida, southwest California
Preferred habitat: Woodlands, swamps, shade trees
Preferred nest site: Under strip of bark low on tree

Clutch size: 4 to 9, usually 5 or 6
Incubation period: 11 to 15 days
Nestling period: 13 to 15 days
Broods per season: 1 or 2
Preferred food: Insects, insect and spider eggs and larvae

WRENS

ONE Native American name for wrens translates as "big noise from little size." Some species of wrens, all insect eaters of the family Troglodytidae, are quite common around human habitation. Providing birdhouses can attract wrens to your property wherever you live (see Chapter 5). Their cheerful songs and feeding habits make them welcome to our gardens whether they deign to visit feeders or not.

Left to right: Carolina wren, winter wren, house wren, Bewick's wren.

WRENS *Troglodytidae*

Winter Wren
(*Troglodytes troglodytes*)

THE male winter wren is a busy fellow. He actually builds several nests and shows them to the female for her approval. She chooses what she thinks is the best of the lot. The male, apparently unwilling to let his labors go in vain, shows the remaining nests to a second female — and gets a second mate. The female lines the nest with soft plant fibers before she lays her eggs.

Size: 3¼"L
Breeding range: Newfoundland west to southern Alaska, south through British Columbia to central California; Washington, northern Idaho; Appalachians south to northern Georgia
Winter range: Southern New England west to southern Illinois, southeast to the Gulf states; Alaska south to southern California, east through central Arizona; southern New Mexico to Texas
Preferred habitat: Coniferous forests to brush piles
Preferred nest site: Tangle near ground

Clutch size: 4 to 7, usually 5 or 6
Incubation period: 14 to 16 days
Nestling period: 14 days
Broods per season: 2
Preferred food: Insects

Troglodytidae

House Wren

(Troglodytes aedon)

THE house wren occurs nearly everywhere in the contiguous United States. These noisy little birds use cavities for nests, whether in trees, birdhouses, or human paraphernalia such as old hats or boots. The male cleans out the nest site (sometimes destroying the nestlings of another bird in the process) and makes a foundation of twigs. Once he has found a suitable site, he begins singing to defend his domain and attract a female. In due time a female appears and inspects nesting sites while he makes trembling movements of wings and tail and performs a flutter flight. Once she has made a selection she will form a cup of grasses, rootlets, feathers, hair.

House wrens are nestlings for sixteen or seventeen days and are brooded for the first three days. During that time the male feeds the female, and the female feeds the young. Both parents feed the young after the female stops brooding them, and both carry off fecal sacs. After one or two weeks of feeding the fledglings, the female leaves to start a new brood; the male continues to feed the young of the current brood.

Size: 4¼"L
Breeding range: New Brunswick west to British Columbia, south to southern California, east to north Texas, Missouri, and South Carolina
Winter range: South Carolina west to California and south to southern Mexico
Preferred habitat: Woods, thickets, farms, gardens
Preferred nest site: Any cavity, including birdhouse
Clutch size: 5 to 8, usually 6 or 7

Incubation period: 12 to 15 days
Nestling period: 12 to 18 days
Broods per season: 2
Preferred food: Insects

Bewick's Wren

(Thryomanes bewickii)

BEWICK'S wren, which resembles the house wren except that it has an unusually long tail, is found in a wide area of the central United States and apparently is expanding its range northward. It nests in a crevice, generally 6 feet up but anywhere from 3 to 25 feet from the ground. In a large structure composed mostly of rubbish and bark, the wren makes a compact, strong, deep cup. The project takes ten days. Bewick's wrens are nestlings for two weeks, fed by both parents. As fledglings they get parental care for two weeks, and then they're on their own.

With Bewick's wren we begin to encounter the real singers. Its song is beautiful, beginning high and rapid and changing to a lower register followed by trills. Its call is a high *chick-chick*, *kut-kut-kut*, or a harsh *spee*.

Size: 4½"L
Breeding range: Southwestern British Columbia southeast to Utah and southwestern Colorado, east to central Pennsylvania, south to Arkansas, west to central Texas and central Mexico, west to California and Baja California

Winter range: Permanent resident throughout most of its breeding range; south to Gulf states
Preferred habitat: Woodlands, thickets, farms, gardens
Preferred nest site: Cavity or birdhouse

Clutch size: 4 to 11, usually 5 to 7
Incubation period: 14 days
Nestling period: 14 days
Broods per season: 2, 3 in the South
Preferred food: Insects

Carolina Wren   *(Thryothorus ludovicianus)*

IN fact, we are unlikely to attract any but the Carolina wren to a feeder. Carolina wrens are found as far north as Nebraska and Massachu-setts, south to Florida and the Gulf Coast. They'll accept peanuts, nutmeats, suet, peanut butter, and cottage cheese. In the garden they occasion-ally snack from bayberries and sweet gum trees. What they really prefer, however, are the insects that constitute about 95 percent of their diet: caterpillars, moths, beetles, grasshoppers, crickets, ants, bees, weevils, and flies.

Carolina wrens nest in a cranny near the ground, usually less than 10 feet up, although 40 feet up has been reported. The nest is a bulky affair of twigs, grasses, and leaves lined with finer materials. It is 6 inches deep and roofed, with a side entrance if it's built in the open. Carolina wrens follow the same incubation schedule as house wrens.

Size: 4¾"L
Breeding range: Southern New England and central New York west to Iowa, south to Gulf coast and Mexico
Winter range: Permanent resident in breeding range

Preferred habitat: Thickets, gardens
Preferred nest site: Any cavity or birdhouse
Clutch size: 4 to 8, usually 5
Incubation period: 12 to 14 days

Nestling period: 12 to 14 days
Broods per season: 2 or 3
Preferred food: Insects

MIMIC THRUSHES

THE mimic thrushes, Mimidae, are related both to true thrushes and to wrens, having some characteristics of each. This strictly American family includes mockingbirds, catbirds, and thrashers.

All of the members of the Mimidae family are accomplished vocalists. Mimidae have seven pairs of intrinsic syringeal muscles, whereas many other species have only one pair.

You can often distinguish between the different mimics by listening carefully to their songs. **Northern mockingbirds,** for example, repeat each phrase three to six times, although individual variations are common. **Gray catbirds'** song is disjointed; they don't repeat their phrases. The **brown thrasher** repeats the phrases of its song twice.

Clockwise from left: California thrasher, gray catbird, brown thrasher, northern mockingbird.

Northern Mockingbird

(Mimus polyglottos)

THE northern mockingbird, the most famous of southern birds, has been gradually expanding its range and is now regularly found as far north as New England. (It is called "northern" to distinguish it from tropical species.) In the east the mockingbird is resident throughout its range, but there is some tendency toward migration in the west. Some shifting toward warmer climates is expected in cold weather. Come winter in the northern reaches of their range, mockingbirds are more likely to be found near the coast (or a well-stocked feeder).

Certainly the songs of the mockingbird can only be described as extraordinary. They produce sounds more usually made by rusty gates, squeaking tricycles, crickets, dogs,

and frogs. One theory is that mockingbirds imitate all those bird songs and make those other remarkable noises by coincidence rather than by mimicry. Believe what you will; I plan to continue loyal to the mimicry theory.

Mockingbirds, by the way, do not seem to mimic the calls of other birds or other sounds if they're busy raising a family or defending territory. Mocking seems to have the status of a hobby for birds not otherwise employed.

Mockingbirds were popular as cage birds before the practice of selling wild birds became illegal. Some were captured as nestlings, others bred in captivity. They have a lot going for them in terms of interesting behavior, even apart from their unquestioned

vocal abilities. They dance and perform somersaults in midair, apparently just for the fun of it. They love to tease dogs and cats, whom they often chase or dive-bomb. They harass snakes and drive intruders of whatever species from the vicinity of their nests. They indulge in free-for-alls among themselves or attack their own images reflected in hubcaps or windows. A mockingbird is seldom without a project.

Mockingbirds are quite happy to live near people. They are especially fond of raisins, but they are adaptable and will eat quite a variety of feeder foods. Such items as American or cottage cheese, apples, berries, currants, nutmeats, peanut butter, suet, and small grains are quite acceptable. A variety of plants will make your yard appealing to them. They like cabbage palmetto, date palms, euonymus, cotoneaster, blueberries, cherries, asparagus, banana, barberry, American beauty berry, camphor tree, sour gum, sumac, mulberry, yucca, cactus, and persimmons. In addition, they eat flies, beetles, ants, caterpillars, grasshoppers, and boll weevils.

Mockingbirds use a feeding territory of about twelve acres and a breeding territory of one or two acres. The male sings, chases, dive-bombs, and displays drooped wings and a loose tail in defense of the breeding territory. Mockingbirds are unusual in that they have fall territory battles to protect their food supply;

however, they usually remain on their summer territory in winter.

The mockingbird picks out an elevated perch from which to sing. From the perch, he jumps and somersaults. After chasing a female, the male performs a loop-flight. Once the female is attracted, they engage in a "dance," facing each other on the ground about a foot apart and hopping from side to side, either in unison or alternately. The male also stretches his wings, moves his spread tail up and down, and coos. He may pick up a twig and run back and forth with it. Maybe that's his promise to help with the nest building.

The nests of mockingbirds

are usually 3 to 10 feet from the ground, in a bush or shrub, but they've been reported from 1 to 30 feet from the ground. The nest is bulky, its outer shell of thorny twigs and its inner construction of leaves, weeds, bark, rootlets, string, rags, cotton, paper, mosses, and hair lined with fine grasses, plant down, horsehair, and rootlets. Both parents build the nest, which is 7 inches in diameter and 4½ inches high, in one to four days. Generally, the male gathers more of the material and the female does more building. Both mockingbird parents feed the nestlings for about twelve days; the fledgling period lasts two to four weeks.

Size: 9"L
Breeding range: Southeastern Maine west to northern California, Mexico to the Gulf states and Florida
Winter range: Permanent resident in most of breeding range; partially migratory in

northernmost parts, retreating southward
Preferred habitat: Deserts and brush to farms to cities
Preferred nest site: Bush or dense tree
Clutch size: 3 to 6, usually 4 or 5

Incubation period: 12 to 14 days
Nestling period: 10 to 13 days
Broods per season: 2, often 3 in the south
Preferred food: Insects, fruits, berries, seeds

Gray Catbird

(Dumetella carolinensis)

GRAY catbirds are likely to be found in thickets, especially near water — they're addicted to bathing — all over the United States except in parts of the Far West and Southwest. Since they prefer shrubby growth, the settlement of the continent has provided additional habitat for them. Consequently, their numbers have increased since the seventeenth century.

Gray catbirds are friendly and relatively tame, but their liking for tangled shrubbery and their ability as ventriloquists makes them hard to observe. Ours are firmly established in the raspberry patch but will come to the feeders for a variety of foods. They, too, dote on raisins, especially steamed ones, and have a taste for soft berries like raspberries, blackberries, grapes, and elderberries. They will eat suet, walnut meats, and cheese at feeders,

and currants, apples, and oranges, as well.

In your garden, they'll be attracted to sand cherries, Boston ivy, autumn and cherry elaeagnus, black gum, blueberries, honeysuckle, alder, sumac, and cherries. In the summertime they eat a lot of insects. One source reports that they eat potato bugs, and I wish they'd get to work on mine. Get out of those raspberries and into the potato patch! They also eat ants, Japanese and other beetles, caterpillars, and grasshoppers.

In courtship the male pursues the female in flight, then bows, raising his plumage, and struts. Catbirds claim a small territory, and the pair

together defends it by song, chases, the body fluff, and the raised wing. The male defends against birds, and the female joins in defense against other intruders. A density of eighty pairs of gray catbirds per one hundred acres has been reported; the typical area for a pair is one to three acres. The ventriloquist catbirds have a mewing call and several notes — *chuck, chatter,* and *kak-kak-kak.*

Gray catbirds like brambly nesting sites, normally not more than 10 feet from the ground (but nests have been reported 20 to 60 feet up). Their concealed nests, built by both parents in five to eight days, are deep, good-sized,

Size: 7¾"L
Breeding range: Nova Scotia west to British Columbia, south to eastern Oregon and northern Utah to east Texas, and south to Gulf states except Florida
Winter range: Coastal areas

from Long Island south to Florida, the Gulf states, and Mexico
Preferred habitat: Wet or dry thickets, hedges
Preferred nest site: Near ground in thicket
Clutch size: 3 to 6, usually 4

Incubation period: 12 to 15 days
Nestling period: 9 to 16 days, usually 10 or 11
Broods per season: 2, sometimes 3
Preferred food: Insects, fruits, berries

scraggly looking, cup-shaped structures made of twigs, vines, cedar bark, weeds, leaves, and grasses lined with rootlets. (Some authorities say that the male merely helps in gathering materials and that all the actual building is done by the female.) Its appearance is quite similar to a mockingbird's nest, but catbirds use fewer manufactured materials. They may use the same nest for second and third broods.

Catbird females brood their young for the first few days. Both parents feed the young; both eat the fecal sacs for the first half of the nine- to sixteen-day nestling period and take them away from the nest for the second half. Fledglings perch near the nest for the first few days and are fed by the parents. Sometimes the male alone feeds them while the female starts a new family.

Mimidae

California Thrasher

(Toxostoma redivivum)

THE California thrasher is smaller than the brown thrasher and has a sickle-shaped bill. This formidable weapon assists him in defending his nest. He's entirely willing to attack people if they venture too close to his family. California thrashers like sumac, elderberry, pyracantha, privet, grapes, berries, and weed seeds. Their insect food includes ants, wasps, bees, termites, and beetles. At the feeder, they're omnivo-rous, devouring dog biscuits, suet, raisins, currants, seeds, grains, and berries with a fine lack of discrimination. They try to drive other birds away and are often successful.

California thrashers are highly aggressive in defending the area near the nest site in their territory of five to ten acres. Like the brown thrasher, they have been known to attack human intruders. California thrashers' nestlings are fed by regurgitation for the first three of their twelve- to fourteen-day stay in the nest.

Size: 10"L
Breeding range: Sacramento Valley in California to northern Baja California
Winter range: Permanent resident in breeding range

Preferred habitat: Chaparral, thickets, parks, gardens
Preferred nest site: Bush
Clutch size: 2 to 4
Incubation period: 14 days
Nestling period: 12 to 14 days

Broods per season: 2
Preferred food: Wild fruits, berries, insects

Brown Thrasher

(Toxostoma rufum)

THE brown thrasher, found only east of the Great Plains, is reddish brown, with a long tail, streaked breast, and long, curved bill. It has a beautiful song that contains much variation and some mimicry, with the phrases occurring in pairs. Although thrashers have as great a fondness for bathing in water as gray catbirds, they prefer to live in dry thickets. They take dust baths, too.

Thrashers are less common around human habitation than the other mimic thrushes. American beauty bush, black gum, currants, blueberries, raspberries, elderberries, and cherries may lure them to your garden. At the feeder, they'll choose cheese, raisins, cracked corn, and walnut meats — which they'll steal from other birds, if they get the chance. Their natural diet includes insects, particularly beetles and caterpillars. They also feed on grubs, cutworms, grasshoppers, wasps, and wireworms.

The songs of the brown thrasher, somewhat subdued during courtship, suggest those of the mockingbird but don't contain as much mimicry. Its call is smaack, and a ventriloquial *chuuuurl* when feeding its young.

To win a mate the brown thrasher struts and trails his tail, and he seeks a new female for the second brood. The pair nests on the ground and up to 14 feet from the ground, preferably in tangled growth. On a loose foundation of thorny twigs, both parents build a large cup of dry leaves, followed by an inner cup of grasses and twigs which is then lined with rootlets. The diameter is 1 foot, and the height 3¾ inches. It takes them five to seven days.

Brown thrasher parents eat the fecal sacs for the first eight days of the eleven-day nestling period and afterward drop them some distance from the nest. Both parents feed the young. The male takes charge during the fledgling period while the female builds a new nest; then he helps her care for the second brood. Be careful: the brown thrasher has a reputation for attacking people who venture too near its nest.

Size: 10"L
Breeding range: Maine west to southern Alberta, south to the Gulf states and Florida
Winter range: Long Island and coastal New Jersey to coastal Virginia, inland from Virginia west to Texas, south to the Gulf states and Florida
Preferred habitat: Dry thickets
Preferred nest site: Thicket
Clutch size: 2 to 6, usually 4
Incubation period: 11 to 14 days
Nestling period: 12 to 13 days
Broods per season: 2, sometimes 3 in the South
Preferred food: Insects, berries, fruits, grains

THE THRUSH FAMILY — ROBINS, THRUSHES, AND BLUEBIRDS

THE thrushes, whose family name is Turdidae, are renowned for singing (the European nightingale is a thrush). They are numerous and found all over the world except in parts of Polynesia. Although they aren't regular visitors at most feeding stations, one or more species will likely be residents of or frequent visitors to your garden. They are included here for that reason.

The **wood and hermit thrushes,** handsome in a more muted way than the robin or bluebird, aren't as likely to be recognized by the novice birder. A better means of identification is their flutelike song: thrushes have a well-deserved reputation as songsters. The wood thrush, which likes a high perch early in the season and a lower one later on, is an excellent singer. The hermit thrush has 150 different songs and is sometimes ventriloquial.

Early on, **bluebirds** captured the affection of people through their beauty and melodious songs and because their habits are largely beneficial to us. They're relatively tame and accept nesting boxes provided for them. Periodically they've suffered large population losses because of severe weather in their breeding range, to which they return fairly early. It's important to provide bluebirds with nesting boxes to maintain their numbers because of the competition for nesting sites from starlings and house sparrows. *See* Chapter 5 for more information on nesting boxes.

There are three major species of bluebirds, all of them very attractive: the eastern, the western, and the mountain. Their habits and diets are quiet similar. In the summertime their food consists mainly of insects; later they add some fruits. All bluebirds have exceptional eyesight and usually hunt from a perch.

Bluebirds can be attracted to stationary feeders. Special enclosed ones that exclude larger birds are available commercially. Their preferred feeder foods are raisins and fruits, but they'll also sample peanuts and peanut butter. Among the insects bluebirds normally eat are grasshoppers, beetles, caterpillars, and crickets, but not many flying insects. In addition to poison ivy and sumac berries, they'll eat blackberries, blueberries, currants, and the berries of asparagus, dogwood, choke-cherry, and huckleberry, to say nothing of elaeagnus, date palm, cotoneaster, Virginia creeper, euonymus, mountain ash, bittersweet, black gum, and camphor tree.

Left to right: Eastern bluebird, mountain bluebird, hermit thrush, Townsend's solitaire, American robin, western bluebird; lower left, wood thrush.

American Robin

(Turdus migratorius)

RECOGNIZED by even the smallest child, the robin is our largest thrush, measuring ten inches. It's found all over the continent throughout spring and summer. Robins migrate by day and often don't go far, though some reach southern Mexico. On the other hand, many robins can be found in the north in winter, when they forsake lawns and move into marshy or swampy places.

The robin is another species that has become more plentiful and has extended its range, especially in the south and west, because of increasing human use of the continent. Its natural habitat is barrens, open woods, or the edges of forests, but it has proved capable of adapting to suburban life. At one time, oddly enough, it was considered a game bird and was shot by the thousands.

We think of robins as relatively tame, but they're not ordinarily feeder birds in the north. If they come to a feeder, they're likely to sample raisins and fruits. Occasionally a robin decides to terrorize feeder society, but it's a passing phenomenon. It's much more likely that your robins will be attracted to the garden for their animal and vegetable food — and to your birdbath. Robins are unabashedly fond of bathing.

Robins eat earthworms, but they eat a lot of other things, too. Ask anyone who has a strawberry patch or a cherry tree. To protect the fruits you want for your own use, provide a substitute — an attractive substitute such as the mulberry. Three other favorites especially attractive to robins are pyracantha, mountain ash, and chinatree berries. Robins tend to like fruit over-ripe, whereupon they proceed to get themselves thoroughly drunk, even falling to the ground in a stupor. This hardly seems the way for a bird to preserve its health — to say nothing of its dignity.

Among robins' other food favorites are the berries of Japanese and American barberry, honeysuckle, juniper, and Russian olive. They'll go after grapes and the fruits of cotoneaster, Virginia creeper, black gum, currant, date palm, persimmon, American beauty berry, bittersweet, euonymus, sumac, madrone, cranberries, cabbage palmetto, camphor tree, pepper tree, and chokecherries. Provide plenty of alternatives and you'll have less competition for your own favorites. Besides earthworms, robins eat flies, caterpillars, spiders, grasshoppers, ants, weevils, and tiny fish, to mention some of their animal food.

Often the earliest singers of the day, robins use a territory of one-third acre to an acre, but territories often overlap, giving rise to chases and fights. Male robins can become quite aggressive after their flocks break up, and they'll fight each other (or their reflection) as if they were gamecocks. They lower their heads and lift their tails, crouching, to attract a female. They are normally, but not necessarily, monogamous.

Robins nest nearly anywhere, often in places very inconvenient for their human neighbors. The nest is cup-shaped, made of twigs cemented with mud and lined with fine grasses or, preferably, domestic animal fur. It is usually 5 to 30 feet from the ground, often in evergreens because robins nest early, when the leaves of deciduous trees haven't yet appeared. It takes two to six days (a high of twenty days has been reported) to build the nest, which has an inside diameter of 4 inches. Male robins alternate between helping the female build the nest and singing to defend the territory. The female smooths the lining with her breast.

Because robins are fond of nesting close to human habitation, young robins sometimes fall victim to cats. The Audubon Society has pointed out, however, that their population remains stable because there are virtually no wild predators in such environs.

Robins are apt to be seen on lawns and in gardens at close range, so you may have a chance some time to observe their parenting behavior. They are among the birds that won't tolerate a cowbird egg in the nest and will chuck it out immediately. A robin sometimes uses the same nest, after refurbishing it, for successive broods. It also uses old nests of other species. Occasionally it builds a new nest on top of an old one. Structures six nests high have been reported. Albino birds sometimes occur among thrushes, including robins. I've seen photographs but never the genuine article; however, my bird-loving aunt in Illinois saw one in her shrubbery and snapped its picture. She reports that the other robins kept their distance.

Robin nestlings usually remain in the nest about thirteen days, though the nestling period may range from nine to sixteen days. Both parents feed the young; both use the drooping-wing tactic to deter people from approaching the nest. They swallow the fecal sacs. The young birds usually remain in the fledgling stage for about two weeks, but the period may last up to four weeks. Though they continue begging, they're mostly the responsibility of the male because the female starts a second brood, sometimes building a new nest (perhaps on top of the old one), sometimes refurbishing the old nest. They gather in large flocks after breeding and engage in the pleasure of mobbing crows.

Size: 10"L
Breeding range: Labrador west to Alaska, south to Mexico, east to Florida
Winter range: Southern New Brunswick west to British Columbia, south to Mexico, the Gulf coast, and Florida
Preferred habitat: Open woods and swamps to cities
Preferred nest site: Tree or nesting box
Clutch size: 2 to 7, usually 3 or 4
Incubation period: 11 to 14 days
Nestling period: 9 to 16 days
Broods per season: 2 or 3
Preferred food: Fruits, earthworms, insects

Muscicapidae

Wood Thrush

(Hylocichla mustelina)

THE wood thrush, somewhat smaller than the robin, has a reddish brown head and a creamy breast with brown spots. At various times it has been called the wood robin, song thrush, and swamp robin. Though its natural habitat is deep woods, it is frequently found near human habitation from the Great Plains east. It winters in southern Mexico to Panama.

If lured to a feeder, the wood thrush will eat suet, peanut butter, raisins, and cornmeal. Some people have fed it honey or syrup. More than half of its food is insects: ants, beetles, spiders, grasshoppers. Like a few other birds, it's said to eat the Colorado potato beetle. It also likes elderberries, mulberries, strawberries, chokecherries, and the fruits of the dogwood and sumac.

The wood thrush defends one-fifth acre to two acres of territory. When it nests near people, its favorite sites are beech trees, grape arbors, dogwoods, and rhododendrons, usually three to twelve feet above the ground, though nests have been reported as high as 50 feet up. Its nest resembles that of the robin — a large cup made of grasses, leaves, weeds, and moss held together with rotted leaves or mud and lined with fine grasses or rootlets. Interestingly, the wood thrush often puts something white on the nest's outer wall.

The female wood thrush, which has a blood red brood patch, broods her nestlings alone. Both parents feed the young during the twelve- to thirteen-day nestling period, swallowing or removing the fecal sacs. Fledglings beg for food, sometimes for as long as thirty-two days.

Size: 7"L
Breeding range: Nova Scotia to South Dakota, south to Texas and east to Florida
Winter range: Coastal Mexico and Central America
Preferred habitat: Moist woods, shade trees

Preferred nest site: Tree or shrub
Clutch size: 2 to 5, usually 3 or 4
Incubation period: 13 to 14 days
Nestling period: 12 to 14 days

Broods per season: 1 or 2, 3 in the South
Preferred food: Insects, fruits

Hermit Thrush *(Catharus guttatus)*

THE hermit thrush, as its name might suggest, is likely to be found alone. It's also known as the American nightingale and the swamp angel. Those names indicate clearly its vocal abilities. The earliest thrush to arrive in spring, it's also the last to leave in the fall, and a few winter in the North. The hermit thrush is smaller than the wood thrush and has a reddish brown tail that contrasts with its dull brown back.

At the feeder, it will be tempted by nutmeats and suet. Its natural food is spiders, ants, crickets, wasps, caterpillars, and beetles as well as wild fruits and berries. To attract it to your yard, plant euonymus, bittersweet, privet, barberry, mountain ash, black gum, and Virginia creeper. Hermit thrushes also like the fruits of wild grape, sumac, and poison ivy. They're unlikely to raid domestic berry patches, seeming to prefer wild berries.

Their preferred nesting sites are on or near the ground in damp, cool woods, preferably near a bog or creek, where they raise one to three broods a year — sometimes including a cowbird, to which it becomes an unintentional foster parent. The compact, cup-shaped nest may be from 3 to 10 feet from the ground. It has a bulky exterior made of twigs, grasses, bark, weeds, ferns, and moss lined with pine needles, rootlets, or porcupine hair. The nestling period is usually twelve to thirteen days, and there may be one to three broods.

Size: 6"L
Breeding range: Labrador west to Alaska, south to high mountains of California and the Rockies, central Minnesota to southern New York, mountains of West Virginia and Maryland

Winter range: New Jersey and southern Ohio south to Gulf states and Florida, southwest British Columbia, California to Texas, south to Mexico and Central America
Preferred habitat: Damp mixed woods, thickets

Preferred nest site: On ground or in small tree
Clutch size: 3 to 6, usually 3 or 4
Incubation period: 12 days
Nestling period: 12 days
Broods per season: 1 to 3
Preferred food: Insects, fruits

BLUEBIRDS

FEW birds bring as much delight as do bluebirds. Their vivid blue plumage, buoyant flight, musical song, and taste for insects make them welcome wherever they choose to nest.

Known by the settlers as blue robins, they thrived in the farmland and orchards of preindustrial America. As the rural landscape has become developed, however, their populations have shrunk. Bluebirds also cannot compete effectively for nesting sites with the introduced English house sparrow and European starling. Nestboxes designed just for bluebirds have reversed this situation, and their populations are now expanding. *See* Chapter 5 for information on providing nesting boxes.

The male usually arrives first in the breeding territory and selects a suitable nesting cavity in early spring. He then will perform a display flight for the smaller female, whose colors are a pale version of his. She builds the nest and incubates the eggs, but both parents take care of the nestlings.

Eastern Bluebird

(Sialia sialis)

EASTERN bluebirds, cavity nesters, make loose nests of fine grass. The female builds the nest in four or five days. The outside diameter varies, but it's usually about 2 inches high. The pair raise two or three broods annually, feeding their nestlings insects for two weeks or so.

Size: 5½"L
Breeding range: Nova Scotia west to southeastern Saskatchewan, south through Texas to Central America, east along the Gulf states to Florida; southeastern Arizona

Winter range: Southern New England west to eastern New Mexico and south to Mexico
Preferred habitat: Orchards, wood edges, farms
Preferred nest site: Cavities in trees, birdhouses

Clutch size: 3 to 7, usually 4 or 5
Incubation period: 13 to 15 days
Nestling period: 15 to 20 days
Broods per season: 2 or 3
Preferred food: Insects, fruits

Western Bluebird

(Sialia mexicana)

THE western bluebird is similar in appearance to the eastern except that its throat is also blue, and the red of the breast extends to the shoul-ders and back. An eater of terrestrial insects when they're available, it switches to fruits, berries, and weed seeds in winter.

Size: 5½"L
Breeding range: Southern British Columbia and western Montana south to west Texas and central Mexico, west to southern California

Winter range: Puget Sound, southern Utah, and southwestern Colorado south
Preferred habitat: Open woods, farms, deserts
Preferred nest site: Cavity in tree, birdhouse
Clutch size: 4 to 8, usually 4 to 6

Incubation period: 13 to 15 days
Nestling period: 14 to 18 days
Broods per season: 2
Preferred food: Insects, fruits, berries, weed seeds

Mountain Bluebird

(Sialia currucoides)

THE mountain bluebird, which occurs in a wide range from the Great Plains west, has a blue back, paler blue breast, and a white belly. The northern part of its range appears to be expanding east-ward. Its food habits are similar to those of other bluebirds, largely insectivorous in summer and turning more to fruits in the winter. The mountain bluebird hovers like a hummingbird while hunting.

Size: 6"L
Breeding range: Southern Yukon east to Manitoba, south through North Dakota and the Black Hills to New Mexico, west through high mountains to southern California

Winter range: Washington to Colorado, south to northern Mexico
Preferred habitat: Open terrain
Preferred nest site: Cavity in tree or cliff, birdhouse

Clutch size: 4 to 8, usually 4 to 6
Incubation period: 14 days
Nestling period: 12 to 14 days
Broods per season: 2
Preferred food: Insects, fruits

Muscicapidae

Townsend's Solitaire

(Myadestes townsendi)

TOWNSEND'S solitaire has been described as looking like a small, rather dull mockingbird. It's a bird of rough, mountainous country, found from the Black Hills west, with a highly regarded song. Flying insects provide the bulk of the solitaire's diet. Although it doesn't regularly come to feeders, it can be enticed to your yard by juniper, cedar, and other plants that have berries in winter.

Size: 6¾"L
Breeding range: Central and eastern Alaska to southwestern Mackenzie in the Northwest Territory, south to central New Mexico and west through the mountains to southern California

Winter range: Southern British Columbia and Alberta south to northern Mexico and central Texas
Preferred habitat: Mountain forests, canyons
Preferred nest site: On the ground

Clutch size: 3 to 5, usually 4
Incubation period: Not known
Nestling period: Not known
Broods per season: 1, perhaps 2
Preferred food: Flying insects

OLD WORLD WARBLERS

THE family Sylviidae is known more familiarly as Old World Warblers. Fairly re-stricted in North America, it's represented by kinglets and gnatcatchers — and gnat-catchers don't come to feeders. Much more sociable are those tiny charmers, the kinglets.

Left to right: Ruby-crowned kniglet, golden-crowned kinglet.

THRUSHES, KINGLETS, AND GNATCATCHERS

Muscicapidae

Golden-Crowned Kinglet

(Regulus satrapa)

THE golden-crowned kinglet, found on both sides of the Great Plains, is easily identi-fied by its black-bordered or-ange or yellow cap. In size it ranges from 3½ to 4 inches; only the hummingbird is smaller. These migratory birds eat insects and insect eggs al-most exclusively but will occa-sionally take suet at a feeder. They nest in conif-erous forests, though in winter they may be seen in deciduous trees and evergreens in the suburbs. They have the reputation of being tame, active little birds. In winter they're likely to be seen with chickadees, brown creepers, and downy woodpeckers.

Size: 3¾"L
Breeding range: In the East, Newfoundland west to Manitoba, south to Wisconsin, south in mountains to North Carolina; in the West, southern Alaska south to California, east to New Mexico; north central Mexico

Winter range: Retreats from northernmost part of breeding range except southern Alaska
Preferred habitat: Conifers, mixed or deciduous woods, thickets
Preferred nest site: Horizontal branch of conifer

Clutch size: 5 to 11, usually 8 or 9
Incubation period: Probably 14 to 15 days
Nestling period: Not known
Broods per season: 2
Preferred food: Insects

Ruby-Crowned Kinglet

(Regulus calendula)

ONLY the male of the ruby-crowned kinglet has a red patch, and it's often concealed. If you see what seems to be a kinglet but color isn't visible on the head and it has white eye-rings, it's a ruby-crowned kinglet. Nearly as tiny as the golden-crowned kinglet, these hardy little birds seem quite undamaged by winter storms. What everyone seems to notice about ruby-crowned kinglets is their songs, which are cheerful and remarkably loud. Insects comprise about nine-tenths of their diet, but they also consume some seeds and berries. At feeders they regularly eat nutmeats, peanuts, and suet. If syrup is available, they'll sample that as well.

Ruby-crowned kinglets, though very active, are tame. During migrations they'll often be found in the company of wood warblers, bluebirds, and nuthatches, but ordinarily they're not as gregarious as the golden-crowned kinglets. Despite the great differences in size, the family is closely related to thrushes. The young birds, however, are not spotted, as juvenile thrushes always are.

Size: 3¾"L
Breeding range: Labrador west to Alaska, south to east central California, east to New Mexico; absent from central section of country; east in Canada from northern Great Lakes to northern Maine

Winter range: Southern New England south and west to Florida, Texas; northern California south to Central America; southeast Oregon, central Nevada
Preferred habitat: Conifers, shrubbery, orchards

Preferred nest site: Coniferous shrub or tree
Clutch size: 5 to 11, usually 7 to 9
Incubation period: 12 days
Nestling period: 12 days
Broods per season: 1
Preferred food: Insects, seeds, fruits

Cedar Waxwing

(Bombycilla cedrorum)

IN the family Bombycillidae there are only two species to consider, the Bohemian waxwing and the cedar waxwing. These lovely brown birds are similar in appearance and easy to recognize because both have crests and broad yellow bands at the tips of their tails. Waxwings are so named because of the waxy red tips on their secondary wings. The Bohemian waxwing is the larger of the two, about the size of a robin; the cedar waxwing is roughly an inch smaller. The cedar waxwing is far more common and can be observed over most of the continent at one time or another. For that reason, we'll confine our discussion to the cedars, sometimes called cherry birds. Remember, though, that the habits of the two are quite similar.

Waxwings are primarily fruit eaters, and except during

the nesting season, they live and travel in flocks numbering from about a dozen to over a hundred. Fortunately, they seem to prefer wild fruit to domesticated — but early cherries may be on their menu if native fruits haven't ripened. Though waxwings will come to feeders for fruit, many bird lovers rely on their gardens to feed these birds. They're fond of juniper berries, mountain ash, dates, cotoneaster, highbush cranberries, currants, the blossoms of blue gum eucalyptus, the buds of Chinese and

Siberian elms, pyracantha berries, sand cherries, Japanese barberry, asparagus, birch seeds and buds, and Hopa crab apples. In our yard, they hang out in the honeysuckle bushes when the berries ripen, and they especially like our old apple tree.

If you'd like them at your feeder as well as in your garden, entice them with raisins, apple slices, orange sections, and sunflower seeds. Both kinds of waxwings eat only a small percentage of animal food, mostly insects, but

Size: 7"L
Breeding range: Newfoundland west to central British Columbia, south to central California, east across U.S. to northern Georgia
Winter range: Southern Nova Scotia west to eastern Nevada,

southern California south to Central America; southern British Columbia to central California
Preferred habitat: Open woods, edges, orchards
Preferred nest site: Horizontal branch on tree; semicolonial

Clutch size: 2 to 6, usually 4 to 5
Inoculation period: 12 to 16 days
Nestling period: 12 to 18 days
Broods per season: 1 or 2
Preferred food: Berries, insects

sometimes will sample meat scraps at a feeder, especially during periods of bad weather.

Waxwings migrate, but their distribution is such that some may be in evidence in most areas at almost any time of the year. Cedar waxwings prefer to nest in orchards, pines, or brushy borders, where they build on a horizontal limb 4 to 50 feet from the ground. Their loose nest is composed of grasses, twigs, string, and yarn lined with rootlets and fine grasses. Both parents build, taking five to seven days. The nest is 4½ to 6 inches in diameter, 3½ to 4½ inches high.

One occasionally hears a hint of criticism about waxwings — it's whispered that they tend to gluttony. In the same breath you'll quite likely hear praise for their gentle behavior and generosity. Their practice of lining up on a branch and passing a fruit or berry from one to the other all down the row and back again is well documented and seems to refute the slur on their appetites. It's possible that these apparently prodigious appetites are noticed primarily because the waxwings travel in flocks, and that the bushes and trees are quickly stripped as result of their numbers rather than individual greediness.

STARLINGS

European Starling

Sturnidae

(Sturnus vulgaris)

ANOTHER immigrant, the European starling had a later start than the house sparrow and was successfully introduced at only one place: Central Park in New York City in 1890. It is now found in abundance from coast to coast and has achieved the dubious distinction of displacing the house sparrow from first place as the bird people best like to despise.

Starlings are members of the Old World family Sturnidae, which includes the talking mynah birds. They're capable mimics, though not as competent in that respect as our native mimic thrushes. When they so choose, they can sing melodiously. But they don't in practice choose that course very often, seeming to delight in hoarse squawks and shrill squeals.

The human quarrel with starlings quickens in proportion to their numbers. Since the species is adaptable, intelligent, and prolific, the battle is growing — especially in cities. Their characteristic gathering into huge flocks is one of the major causes of our dismay. Much as a city dweller might admire their spectacular aerial maneuvers, when thousands of them roost on a bridge or a building, they pose a problem. Similarly, when enormous flocks descend on an orchard or grain field, their very numbers assure a great amount of damage in a short time. Those whose crops disappear are unlikely to think kindly of the marauders even if reminded that starlings launch similar attacks on insect pests, especially when there's a virulent infestation.

The starling has been described as "casually omnivorous." Beetles, fruits, berries, nuts, seeds, meat, suet, bread, carrion — you name it, a starling will eat it. A friend swears that they steal sneakers for the sheer joy of ingesting them. They're usually on any list you see of feeder pests, charged with tearing suet bags in their greediness and chasing more desirable birds from the feeder.

Unless you live in a city plagued with starlings, you can cope with them without resorting to a reign of terror. Starlings will be among the first to discover a new feeding station, but they aren't attracted to swinging feeders, which will discourage competition with some of the smaller birds you especially like. And try to keep an open mind. In the spring, a starling's diet is as high as 90 percent animal. That's insects. Unless you're the operator of a feedlot (one in Caldwell, Idaho, reports that their resident starling flock devours fifteen to twenty tons of potatoes a day), a harassed orchardist, or an equally harassed city dweller, you can probably learn to live with starlings.

In courtship, starlings wing-wave and chase each other. They save the interesting stuff for nest building, where it counts. Their nesting behavior is apt to cause some problems. They prefer to be cavity nesters and may decide to take over all your birdhouses if you don't take great care about the size of the entrance holes. What may be more annoying to you personally is their fondness for taking up residence in the ivy growing on your house. A chimney area will claim their interest in winter. Their nonmelodious songs may drive you frantic.

Being cavity nesters, they nest in crevices on buildings or in trees, usually 10 to 30 feet from the ground; however, nests have been reported from 2 to 60 feet up. The male appears on the site first, clears out anything that might be in the cavity, and brings dead leaves, bark, moss, lichens, green leaves, and tree flowers to the chosen place. Then the female comes along and throws out all he's brought and in a few days builds a nest to her satisfaction — mostly of grasses, although weeds, leaves, and cloth may be added. The lining is of fine grasses and feathers. Some observers report that it's built in a slovenly manner and is carelessly kept. (A bad reputation taints everything.) In any case, she builds a new one for each brood. Nests may be solitary or in colonies.

Nestling starlings are fed by their parents for sixteen to twenty-three days but remain in the fledgling stage only four to eight days. During that period they beg incessantly, managing to look remarkably helpless despite their nearly adult size.

Size: 6"L
Breeding range: Southern Canada, throughout United States except southernmost Texas to northern Mexico
Winter range: Resident within breeding range

Preferred habitat: Farmlands to cities
Preferred nest site: Cavity in tree, birdhouse
Clutch size: 2 to 8, usually 4 to 6

Incubation period: 11 to 13 days
Nestling period: 16 to 22 days
Broods per season: 2 to 3
Preferred food: Insects, seeds, fruits, grains

Red-Eyed Vireo

(*Vireo olivaceus*)

RED-EYED vireos defend one to two acres of forest habitat through song, chases, tail-fanning, and crest-erecting. The males chase females, but the rest of their show is merely the tail-fan and sway. The females build nests noted for their beauty. In four or five days they select the site, gather materials, and build a nest with a 2-inch inside diameter of bark, grasses, and spider egg cases and lined with plant down.

The female red-eyed vireo broods her nestlings until the sixth of the ten days they remain in the nest. She eats the fecal sacs at first and later carries them away. The fledgling stage lasts two or three weeks, during which the family roams in search of food.

Size: 5"L
Breeding range: Nova Scotia west to Northwest Territories and British Columbia, southeast to Texas, Gulf states, and Florida

Winter range: South America
Preferred habitat: Woodlands or shade trees
Preferred nest site: Bush, tree
Clutch size: 2 to 5
Incubation period: 12 to 14 days

Nestling period: 10 to 12 days
Broods per season: 1 or 2
Preferred food: Insects

Ploceidae

House Sparrow

(Passer domesticus)

WHEN is a sparrow not a sparrow? When it's a weaver finch. The weaver finch family, Ploceidae, contains the house, or English, sparrow. One of the often despised birds, the house sparrow, like starlings and rock doves (the common pigeon of urban areas), is an immigrant. Brought from the Old World, where they had long been accustomed to city life, house sparrows were first introduced to this continent during the 1850s, and they proved extraordinarily adaptable and prolific. Our native birds had not yet adjusted to urban conditions when the house sparrows arrived and usurped the place that one or more native species might eventually have taken as their own.

They multiplied and multiplied, producing two or three (some birders swear they sometimes produced as many

as five) broods a year. Diseases, weather, and predators eventually took their toll, but it is thought that the coming of motor vehicles did more to help stabilize their numbers because the gradual disappearance of horses from our streets meant a loss of food to the birds. They fed not only on grain spills but also on the waste grain in manure, as well. At the peak of their population explosion, it's estimated that they were the most abundant bird in their North

American range; moreover, their numbers in some areas were *double* those of all other birds together.

House sparrows can also be a real problem for bluebirds in that they peck the young and adults to death inside their nesting boxes. To prevent their taking up residence in a bluebird nestbox, some owners open the boxes and remove house sparrow eggs — repeatedly, if necessary.

There's no need to encourage house sparrows to come to

Size: 5¼"L
Breeding range: Throughout inhabited United States and Canada
Winter range: Permanent resident
Preferred habitat: Farms, villages, cities

Preferred nest site: Cavities, crevices in buildings, trees, vines, birdhouses
Clutch size: 3 to 7, usually 5
Incubation period: 12 to 13 days

Nestling period: 13 to 18 days
Broods per season: 2, sometimes 3 or 4
Preferred food: Insects, seeds, garbage

feeding stations. They're opportunists, and they'll be among the first to discover the existence of a new source of food. But you don't have to poison, trap, or shoot them. They do very little harm that isn't balanced by some of their more acceptable habits. Yes, they're fond of grain and may help themselves where cereals are grown. They may feed on newly sprouted garden vegetables. However, they're also fond of the seeds of crabgrass, dandelion, and ragweed. Sudden infestations of insects — Japanese beetles, for example — will attract them.

House sparrows may nest in places you don't want them. The ivy climbing up your chimney is a favorite spot, and they'll use it for warmth and protection in winter as readily as for a nesting site during the breeding period. They'll steal from other birds at the feeder, but they prefer ground feeding or stable feeders. A hanging feeder will discourage them from gobbling up all the seeds and suet you set out.

Male house sparrows become as quarrelsome as robins at mating time. They wing-stretch and wing-quiver and bob in a crouch. The female opens her bill and lunges at the male, looking rather as if she intended him serious bodily harm. He evidently knows better.

House sparrows become sociable again during nesting; it isn't unusual to find four or five nests in close proximity in one small tree. Truly monogamous and quite sedentary, these homebodies use their nests as year-round homes. They defend only the nest itself, with the *churr* call and head extended forward. House sparrows are chirpers *(chirp, cer-eep)* rather than singers; in courtship they utter a soft, clear warble. Youngsters have a subsong and can be taught to sing, so it's assumed that the chirping is an evolutionary advance, not a failing.

The house sparrow will nest in a cavity in a tree, a crevice in a building, or elsewhere,

usually from 10 to 30 feet from the ground. The nest resembles a huge ball of grasses, weeds, and assorted trash, and has an opening on the side. It's large, bulky, and loosely constructed. Both parents build (some observers insist that the female alone builds), lining the nest with feathers, hair, or string. This lined part is at the center and is relatively compact. Nests vary in size, depending on the area of the site. Sometimes, like wrens and blackbirds, they take over the nest of another species, destroying the eggs or young.

Female house sparrows develop brood patches. Both parents feed the young for their fifteen- to seventeen-day nestling stage, and about 60 percent of the time they have help from an unmated bird. The staple of their diet is insects, and at first the young are fed by regurgitation. Their chief predators are crows and jays. The fledgling stage lasts about a week, during which the young beg, quivering their wings.

WOOD WARBLERS

WARBLERS are often compared to butterflies because of their flitting habits and small size. There are tremendous numbers of warblers, and their habitual activities make precise observations arduous and often frustrating. Bird lovers who do most of their watching from the kitchen window will not have much chance to become familiar with warblers.

There has been some change in nomenclature in the warbler family, now part of the large family Emberizidae, which includes sparrows, blackbirds, and tanagers. For a while they were called Parulidae, before that Compsothlypidae, and before that, Mniotilidae. If you're looking for a particular species, especially in order books, use the name warblers first.

The warbler family is large, containing more than fifty species in North America. Primarily insectivorous and arboreal, they seldom come to feeders. One of the few exceptions is the yellow-rumped (myrtle) warbler, which readily accepts feeders. At migration times, though, you may see various warblers in your garden. Some will be attracted to water; others will visit feeding stations. Because there are so many of them, we'll concentrate mainly on those most frequently seen around homes.

It has been estimated that 40 percent of the world's forests will disappear in the next twenty years and that this loss of habitat will result in destruction of 400,000 to 500,000 species of living things. That's a fifth of the total, a startling number. Not all of that total is presently considered endangered, but one of the species is very well known, having had much attention devoted to it because its needs are incredibly specific. Few of us have ever seen Kirtland's warbler, but we've been made aware of its plight.

In the summertime Kirtland's warblers inhabit six counties in Michigan. They winter in the Bahamas. The population has been steadily declining, although individuals have recently been reported in Quebec. Whether the birds sighted were strays or pioneers expanding their range isn't known. Apart from loss of habitat (they like Jack pines, only of a certain size), Kirtland's warblers are badly afflicted by a related problem: parasitism by cowbirds. Cowbirds expand their range, moving in when farmers do, after loggers have removed the pine trees the warblers need.

Kirtland's warblers are not feeder birds, and I call attention to them only because each of the numerous warblers occupies a certain environmental niche. It is often so well adapted to that niche that it's capable of utilizing no other. If that environment changes, some birds are doomed to suffer and may disappear entirely. Others may become more abundant and yet remain little known to us.

Left to right: American redstart, orange-crowned warbler, yellow-breasted chat, yellow-rumped warbler, common yellowthroat, black-and-white warbler, yellow warbler.

Black-and-White Warbler

(Mniotilta varia)

THE black-and-white warbler is found from the Great Plains east and occasionally visits feeders for suet. Like the brown creeper, it spends its time climbing trees and digging insects from the nooks and crannies in the bark. Although the black-and-white warbler forages in trees, the female nests on the ground. The nest is little more than a depression in leaves. When the chicks are eight to twelve days old, they climb a tree and receive food from the parents.

This warbler devours insects such as gypsy moths and tent caterpillars that can be destructive to trees. Unfortunately its numbers have decreased in recent years, possibly as a result of destruction of its winter habitat.

Size: 4½"L
Breeding range: Southern Canada through southwestern Northwest Territories to northeastern British Columbia, south through eastern Montana to central Texas and Louisiana, northeast to South Carolina

Winter range: Florida, southern Texas, Gulf states coastally south through the tropics to South America; Baja California south
Preferred habitat: Woodlands
Preferred nest site: Stump, on ground
Clutch size: 4 to 5
Incubation period: 13 days
Nestling period: 11 to 12 days
Broods per season: 1
Preferred food: Insects

Orange-Crowned Warbler (*Vermivora celata*)

THE orange-crowned warbler breeds in the West and Northwest, then winters in the South, though it has been found as far north as Massachusetts. It will occasionally visit feeders for suet and nutmeats.

This particular bird is hard to distinguish from certain other warblers, such as the Tennessee warbler *(V. peregrina)*, since its orange crown can be hard to see. Like many other warblers, it has a cheerful song, nests on the ground, and is a useful devourer of insects.

Size: 4½–5½"L
Breeding range: North-central Alaska across Canada to just north of the Great Lakes, western Montana south to west Texas, west to Baja California

Winter range: Southern United States to Guatemala
Preferred habitat: Open woodlands, brush
Preferred nest site: On ground or in low shrub

Clutch size: 4 to 6, usually 4 or 5
Incubation period: 12 to 14 days
Nestling period: 8 to 10 days
Broods per season: Not known
Preferred food: Insects

Yellow Warbler

(Dendroica petechia)

THE yellow warbler, found throughout the contiguous United States, has been known to erect six-story nests just to cover up cowbird eggs. Unfortunately, yellow warblers don't seem to be able to distinguish their own nestlings. If the cowbird egg happens to hatch, the tiny foster parents rear it. Yellow warblers may need up to twenty acres of territory, but in a favorable habitat sixty-eight pairs have been found in a square mile.

One source says that the only warbler likely to nest in a garden is the yellow warbler. They especially like willows and alders, wild roses and grapes for their site, preferably about 10 feet from the ground.

Size: 4¼"L
Breeding range: Canada and Alaska south to northern south America except Texas, Louisiana, southern Gulf states; Florida Keys
Winter range: Southern California, southwest Arizona south to Central and South America
Preferred habitat: Willows, farmlands, dense shrubbery, gardens
Preferred nest site: Shrub, tree
Clutch size: 3 to 5, usually 4 or 5

Incubation period: 10 to 11 days
Nestling period: 9 to 12 days
Broods per season: 1
Preferred food: Insects

Yellow-Rumped Warbler *(Dendroica coronata)*

THE most abundant warbler is the yellow-rumped warbler. It has two subspecies. The eastern form, the myrtle warbler, may be identified by its white throat, yellow rump, and yellow patches on the sides of the breast. It breeds in northern coniferous forests and winters from New England to the South, Southwest, and California coast. Primarily an insect eater, it also eats wax myrtle berries, euonymus, almonds, butternuts, red cedar and poison ivy berries, and sumac and goldenrod seeds. At feeders it consumes bread, cornmeal, peanut butter, nutmeats, sunflower seeds, and suet.

The western form, with a yellow throat, is Audubon's warbler, found west of the Great Plains. A bird lover in South Dakota had one stray that spent the winter at her feeder. She made pie crust for it.

Yellow-rumped warblers have a liquid song. They nest from 5 to 30 feet from the ground in a spruce or fir tree. The male brings a little material for building, but the construction falls to the female, who uses dried grasses, weeds, and bark bound by spiderwebs and lined with hair or plant down. It takes her about a week, and the nestling stage of yellow-rumped warblers lasts twelve to fourteen days.

Size: 4¾"L
Breeding range: Alaska and Canada south to mountains of Massachusetts, New York, and Pennsylvania
Winter range: Southern New England southwest to central California and south to Central America; along Pacific coast from southwestern British Columbia south
Preferred habitat: Forests to gardens
Preferred nest site: Conifer
Clutch size: 3 to 5, usually 4 or 5
Incubation period: 12 to 13 days
Nestling period: 12 to 14 days
Broods per season: 1 or 2
Preferred food: Insects, berries

Common Yellowthroat *(Geothlypis trichas)*

COMMON yellowthroats defend from a half acre to two acres by singing vocal duets from exposed perches, chases, short flights around the boundaries, wing-flicks, and song-flights. The latter consist of flying at an angle for twenty-five to one hundred feet, giving sharp call notes. These birds defend until the nestling stage (after the eggs have hatched) and then again for the second brood.

The common yellowthroat nests in the brush of open fields, both on the ground or to a height of about 2 feet. Made of grasses and leaves and lined with grasses and hair, the nest has an inside diameter measuring 1¼ inches.

Both parents of common yellowthroats, which are frequently parasitized by cowbirds, feed their nestlings for eight or nine days. Sometimes the male gives the food to the female, which then gives it to the nestlings. The fledgling stage lasts two or three weeks and starts before the young can fly. After three days they can fly a little, and after twelve days they're experts. They leave the parents after about twenty days.

Size: 4¼"L
Breeding range: Central Canada south to southern Mexico
Winter range: Southern Maryland south to Texas, west to California, south to Central America
Preferred habitat: Swamps, thickets
Preferred nest site: Weeds, shrubs
Clutch size: 3 to 6, usually 4
Incubation period: 11 to 13 days
Nestling period: 9 to 10 days
Broods per season: 1 or 2
Preferred food: Insects

Yellow-Breasted Chat *(Icteria virens)*
American Redstart *(Setophaga ruticilla)*

THE yellow-breasted chat, or wood warbler, is the largest warbler (about 7½ inches) and is regarded as eccentric, clownish, and altogether unwarblerlike. It may be attracted by blackberries, blueberries, elderberries, dogwood, and sumac. It's found all over the contiguous states.

Among the warblers you may see is the American redstart, unmistakable with its orange patches in wings and tail. It is likely to come to a birdbath. In the Northeast, it may be joined by the Blackburnian warbler.

Yellow-Breasted Chat

Size: 6¼"L
Breeding range: New Hampshire west to southern British Columbia, south to northern Baja California, northern Mexico, Texas to northern Florida
Winter range: Mexico and Central America to Panama
Preferred habitat: Thickets, briars, brushy pastures
Preferred nest site: Thicket
Clutch size: 3 to 5, usually 4
Incubation period: 8 days
Nestling period: Not known
Broods per season: 1
Preferred food: Insects, berries

American Redstart

Size: 4½"L
Breeding range: Southern Labrador west to Northwest Territories, southeast Alaska south to Oregon, southeast to eastern Texas and Louisiana to mountains of Georgia
Winter range: Mexico to Ecuador, south Florida
Preferred habitat: Deciduous woodlands, swamps, shrubbery, shade trees
Preferred nest site: Bush or tree
Clutch size: 3 to 5, usually 4
Incubation period: 12 to 14 days
Nestling period: 8 to 9 days
Broods per season: 1
Preferred food: Insects

Bobolink
Eastern Meadowlark
Western Meadowlark

(Dolichonyx oryzivorus)

(Sturnella magna)

(Sturnella neglecta)

MEADOWLARKS and bobolinks are tuneful members of the blackbird family who rarely come to feeders. They nest in meadows, hiding the nest under a canopy of tall grass, carefully constructing it out of grasses and lining it with finer materials. Primarily insect eaters, they can occasionally be attracted to grain scattered on the ground.

Bobolink

Size: 6"L
Breeding range: Nova Scotia west to southern British Columbia, south through eastern Washington to Colorado, east to coastal Maryland
Winter range: South America
Preferred habitat: Open grasslands
Preferred nest site: On ground in dense vegetation

Clutch size: 4 to 7, usually 5 or 6
Incubation period: 13 days
Nestling period: 10 to 14 days
Broods per season: 1
Preferred food: Insects, seeds

Eastern Meadowlark

Size: 9"L
Breeding range: New Brunswick west to Ontario, south through Nebraska to Texas and eastern Arizona to northern Mexico, east to Florida
Winter range: Central New England to Great Lakes and south
Preferred habitat: Meadows
Preferred nest site: On ground in vegetation
Clutch size: 2 to 6, usually 3 to 5

Incubation period: 13 to 15 days
Nestling period: 10 to 12 days
Broods per season: 2
Preferred food: Insects, seeds, grains

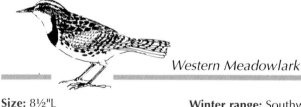

Western Meadowlark

Size: 8½"L
Breeding range: Central British Columbia to southern Ontario, south to northeastern Louisiana, west through Texas and northern Mexico to Baja California
Winter range: Southwest British Columbia south and east to Gulf states, south to central Mexico
Preferred habitat: Meadows, fields, prairies
Preferred nest site: On ground in grass

Clutch size: 3 to 7
Incubation period: 13 to 15 days
Nestling period: About 12 days
Broods per season: 2
Preferred food: Insects, grains

BLACKBIRDS

OPINION is divided regarding these two handsome blackbirds. In northern areas the song of redwings, as they return in large flocks to their breeding range, is a welcome promise of spring. On their winter range, however, those large flocks can quickly strip a farm field of valuable grain. In summer, blackbirds generally concentrate on insects, many of them harmful.

The male and female roles of these species are carefully delineated, as is usually the case where their plumage is so different. The male arrives first in the breeding range and chooses a territory, usually not far from that of another male. When the female comes a few weeks later, the male displays before her. She chooses the nest site; he defends the site against other males. The female weaves a basket-like nest to hold the three to five eggs, and she does the incubating. If there is a surplus of females, male redwings will have several mates.

Left to right: Yellow-headed blackbird pair, red-winged blackbird pair.

Red-Winged Blackbird

(Agelaius phoeniceus)

RED-WINGED blackbirds, found all over the continental United States, are handsome birds and easy to recognize. They are extremely gregarious, nesting in loose colonies of anywhere from a few to several hundred pairs. Despite this sociability, the males vigorously defend the cubic yards around their nests with song, bill-tilt, song-flight, and tail-flick. Against intruders of other species, whether bird or mammal, red-winged blackbirds are ruthless. Many birds dive at humans, but the redwings will actually strike, banding together for the purpose.

The female selects the nesting site, which is low and usually in a marshy area or near water. The cup-shaped nest is made of long strands of rushes, grasses, and moss, plastered with mud and bound to the surrounding vegetation

with milkweed fibers and lined with fine grasses. She spends three to six days in gathering material and construction. The nest is generally 3 to 8 feet from the ground, often suspended, usually in a shrub. Some nests have been observed in trees as much as 30 feet from the ground; others are on the ground.

It is mostly the female red-winged blackbird that cares for the nestlings for their eleven days of nest life. She will wing-flap — one or both wings — at any person who approaches the nest. Both parents feed during the fledgling period, which starts with the nestlings crawling out of the nest to perch nearby. That stage lasts a week to ten days.

Polygamy and polyandry are more common to game birds than to songbirds, but redwings are exceptions. Male red-winged blackbirds average

Size: 7¼"L
Breeding range: Newfoundland west to Alaska, south to Mexico, and east to Florida
Winter range: Southern New England west to British Columbia, south to South America

Preferred habitat: Fields, marshes, woods edges
Preferred nest site: Attached to reeds in marsh; loose colonies
Clutch size: 3 to 5, usually 3 or 4
Incubation period: 10 to 12 days

Nestling period: 10 to 11 days
Broods per season: 2 or 3
Preferred food: Insects, seeds, grains

three mates each. They perform song-flights, crouch, and chase their chosen females, with each of whom they spend about three weeks.

After the nesting season, redwings gather in flocks and use communal roosts, often joined by grackles and starlings. Because of their numbers, they can be a nuisance to farmers. Sometimes, especially in the southern states, the roosting grounds contain millions of birds. Generally their diet consists of about one quarter insects and three-quarters seeds and grains, including barley, rice, white ash, American beech, and cabbage palmetto. In summer as much as half of their diet may consist of ants, beetles, grasshoppers, snails, and spiders. They prefer to eat on the ground.

Though red-winged blackbirds move southward for the winter, they don't go far — or for very long — and a few will stay though the winter and visit feeding stations. This can be a mixed blessing since they're large, aggressive, and possessed of vigorous appetites. They like practically everything: cracked corn, hemp, millet, raisins, apples, dog food (dry or wet), and peanuts.

WOOD WARBLERS, SPARROWS, BLACKBIRDS, AND TANAGERS *Emberizidae*

Yellow-Headed Blackbird (*Xanthocephalus xanthocephalus*)

THE flashy yellow-headed blackbird is not considered a feeder bird, but grain scattered on the ground may attract it. Its diet is one-third insects, the rest grain and seeds. It's generally found from the Great Plains west. Because it and the red-winged blackbird tend to occur in large numbers, their welcome at feeding stations is doubtful.

Size: 8½"L
Breeding range: Central British Columbia east to southern Manitoba, northwest Ohio southwest to New Mexico and northern Baja California
Winter range: Southwestern United States and Mexico

Preferred habitat: Fields, marshes, farms
Preferred nest site: Attached to cattails in marshes; in loose colony
Clutch size: 3 to 5, usually 4
Incubation period: 12 to 13 days

Nestling period: 12 days
Broods per season: 1
Preferred food: Insects, grains, seeds

243

ORIOLES

Finally we come to the royal branch of the Icterinae subfamily, the orioles. American orioles are not related to European orioles. Their common name was the result of error, but it is now too firmly entrenched to change it. Another source of confusion has arisen regarding Baltimore and Bullock's orioles. They were considered the eastern and western race, respectively, of the same species — the **northern oriole** — but ornithologists may split them again into two distinct species.

Baltimore and **orchard orioles** are found from the Great Plains east; **Bullock's oriole**, from the Great Plains west. As sometimes happens, birds wander out of their accustomed range. Verified reports of both orchard and northern (Baltimore) orioles have been recorded in California.

Certain other orioles occur only in extremely limited ranges in south Texas, southern Florida, southern Arizona, and southwestern California. All orioles normally winter south of the United States. They aren't common feeder birds during cold weather, but if for some reason they stay in their nesting areas, they will come readily to feeders.

All oriole males have a loud, pleasing, almost metallic whistle and are useful insect eaters.

Left to right: Orchard oriole and northern orioles — center, Bullock's; right, Baltimore.

Orchard Oriole   *(Icterus spurius)*

ORCHARD orioles, found in the same range as northern (Baltimore) orioles, are sometimes overlooked because their colors are more muted. Novices may even think them the same species. Where the northern oriole is brilliant orange, the orchard oriole is chestnut; its tail is entirely black, it has less white in the wing, and it's slightly smaller. Orchard orioles do nothing to defend territory and are even willing to share a tree with another pair. The male has an attractive song similar to that of a bobolink or fox sparrow. When courting, he simply bows before the female and sings vigorously. Maybe they're both too frazzled with the whole idea of getting the cycle finished up by July, when they migrate south, to waste time on preliminaries.

As the name implies, orchard orioles frequently nest in orchards. Their nest doesn't swing free the way the northern oriole's does. The female builds the large, rounded nest, with a constructed rim, of weeds, fibers, and dried grasses. Their stay is quite brief: they raise one brood and then they're gone, usually by the end of July.

Their diet, fortunately, is mostly insects, although they sometimes eat mulberries. They seldom come to feeders but have sometimes been attracted by bread spread with jelly.

Size: 6"L
Breeding range: Eastern Massachusetts west to south-central Saskatchewan, south to northern Mexico, the Gulf states, and northern Florida

Winter range: Mexico to northern South America
Preferred habitat: Orchards, farmlands, towns
Preferred nest site: Branch in dense foliage

Clutch size: 3 to 7, usually 4 or 5
Incubation period: 12 to 14 days
Nestling period: 11 to 14 days
Broods per season: 1
Preferred food: Insects, fruits

Northern Oriole

(Icterus galbula)

NORTHERN orioles will often come to a feeder for berries or orange slices, nutmeats, or suet. They also like syrup, and some learn to use hummingbird feeders. They're among the birds that appreciate provided nesting materials such as thread, string, or yarn. But keep the pieces short (six to eight inches) so the birds don't get entangled in loops and strangle themselves.

The northern oriole gives six or seven loud whistles in announcing his territory and a loud, harsh chatter as an alarm call. In courtship, monogamous northern orioles droop their wings and tails, fan them, bowing their heads and whistling, then flutter in the air.

Female northern orioles select the site for their nests, usually on a tree branch hang-

ing over a clearing. She takes five or six days to build a hanging, gourd-shaped nest about 5 inches long. Its outer shell is tightly woven of weed stalks and inner bark and perhaps string, and the bottom is filled with plant down or hair. The female bounces in the nest, using her breast to shape it.

Northern orioles feed insects to their young, at first by regurgitation. The nestling period is about two weeks, just as it is for orchard orioles.

About half the diet of northern orioles is animal and includes gypsy moth larvae, grasshoppers, tent caterpillars,

and leaf beetles. Orioles are also fond of soft fruits and berries, green peas, mountain ash, and the seeds of garden flowers such as hollyhocks and sunflowers. We know it's time to harvest the pears when the orioles begin spending their time in the tree. Fall migration to the tropics begins in September, though scattered individuals may remain until October.

Size: 6½"L
Breeding range: Nova Scotia west to British Columbia, south to southern California and Mexico, east to North Carolina

Winter range: Mexico to northern South America
Preferred habitat: Shade trees and woods edges
Preferred nest site: High branch of deciduous tree

Clutch size: 3 to 6, usually 4 or 5
Incubation period: 12 to 14 days
Nestling period: 11 to 14 days
Broods per season: 1
Preferred food: Insects, fruits

Brewer's Blackbird *(Euphagus cyanocephalus)*

BREWER'S blackbird, a bird of fields and farmyards, occurs from the Great Plains west. It resembles a common grackle in appearance but is smaller, with a shorter tail. It often travels in flocks with red-winged blackbirds or brown-headed cowbirds. Although Brewer's is fond of bread, it isn't considered a feeder bird.

The male is very aggressive in defending his mate as she constructs a nest of interlaced twigs, grass, rootlets, and mud on a low branch of a shrub or conifer tree. She then incubates the eggs by herself for twelve to fourteen days.

Size: 8"L
Breeding range: Central British Columbia east to southwestern Ontario, south through Michigan to New Mexico and west to southern California

Winter range: Southern British Columbia, southeast through South Carolina, Florida, and Mexico
Preferred habitat: Fields, ranches, farms
Preferred nest site: On ground,

bush, tree; loose colonies
Clutch size: 4 to 8, usually 4 to 6
Incubation period: 12 to 14 days
Nestling period: 13 days
Broods per season: 1
Preferred food: Insects, seeds

Common Grackle   (*Quiscalus quiscula*)

GRACKLES are larger than robins, about the size of blue jays. Superficially they resemble crows, but their smaller size and keeled tail make them readily identifiable. Their ability to adapt to changes wrought by civilization has enabled them to increase in numbers despite changing conditions. Once largely birds of the marshes, they've now invaded farmland, suburbs, parks, and urban areas. They're often among the first to discover a new feeding station, and they're quite capable of taking over. Competent thieves themselves, they seem to object violently when they're victimized by other birds. At the feeder they'll take whatever is on the menu, but they're particularly partial to corn, suet, nutmeats, sunflower seeds, and wheat. If offered hard bread products, they're apt to take them to the birdbath to dunk before eating them.

Purple and bronzed grackles, which hybridize freely wherever their ranges overlap, are now considered subspecies of the common grackle rather than separate species. The purple or blue-green subspecies is more likely to be found in the South and East; the bronzed, west of the Alleghenies. They're not usually found west of the Great Plains.

Among landscape plants that appeal especially to grackles are American beech and cabbage palmetto. Corn and wheat are their favorite cultivated grains, which they start raiding in July and August. After harvest, they primarily rely on waste grain and weed seeds. Most of them migrate south and sometimes hang around barnyards in flocks to steal grain from domestic livestock.

Grackles eat animal matter too — house sparrows, an occasional egg or nestling, or sometimes a field mouse. In the early spring they hunt small crustaceans and insects in marshes, and when plowing starts, they have a (literal) field day with the turned-up insects. Once it was thought that they hung around after seeds sprouted just to wreak havoc on the young plants (that's what the wrong image can do for you), but it has since been determined that they're actually after cutworms. They feed their nestlings exclusively on

insects and they eat more Japanese beetle grubs than any other bird — proving they're not totally devoid of usefulness after all.

In their colonies, common grackles defend four to eight square yards at the nest site by perch-taking. Their courtship song is a squeaky, shrill wheeze. The males approach the females with heads down and bills tilted, fanning wings and tail and ruffling the feathers of neck and back. The female wing-quivers in acceptance. The male then leaves her while she's incubating the eggs and finds another mate, which is probably just fine for the species but seems a trifle hard on the female.

Common grackles are somewhat colonial in nesting habits. They are partial to evergreens, 3 to 30 feet from the ground. The female builds the loose, bulky nest, which has an inside diameter of 4 inches, an outside diameter of 7 to 9 inches, and is 5 to 8 inches high. Sometimes she builds on an osprey nest. Her own is well made, of twigs, coarse grasses, and seaweed lined and cemented with mud, then lined again with soft grasses and feathers. It may take the parents up to six weeks to gather materials, but the female does the building in about five days.

Grackles eat or carry away the fecal sacs of nestlings. The male grackle helps the female feed nestlings for the twelve days they stay in the nest only if he hasn't found a new mate. There's some question of whether there's any fledgling stage at all for grackles; if there is, it's very brief and one or both parents feed the juveniles.

Size: 10–12"L
Breeding range: Newfoundland west to southern Mackenzie (Northwest Territory), south to New Mexico, the Gulf states, and Florida
Winter range: Central New England and New Jersey west to Kansas, south to eastern Texas, the Gulf states, and Florida
Preferred habitat: Croplands, lawns
Preferred nest site: Tree, bush; often in colonies
Clutch size: 4 to 6, usually 5

Incubation period: 11 to 13 days
Nestling period: 14 to 20 days
Broods per season: 1
Preferred food: Insects, fruits, grains
Preferred feeder food: Corn, suet, sunflower seeds

249

Brown-Headed Cowbird *(Molothrus ater)*

I CAN'T think of a single common bird with a worse reputation than the cowbird — and you know how it is once anybody gets a bad reputation. All kinds of villainy, justified or not, is attributed to cowbirds. The source of the outrage is that cowbirds choose to leave the care of their offspring, from egg to adult, to some other species.

Here's what happens. The male cowbird — about robin-sized, black with a brown head — mates with the gray female; when she feels the urge to lay an egg, she scouts around for an available nest. She lays one to six white eggs splotched with brown, not all in the same nest, escaping notice by depositing her eggs when the host is absent — often just before dawn. Something like 195 species have been unwitting recipients of her eggs, including such unlikely victims as hawks and gulls (the efforts failed in those cases) — though usually the host eggs are smaller than the cowbird's. The most usual targets are nests of vireos, warblers, sparrows, and flycatchers.

All's well (for the cowbirds) if the female host has finished laying and is about to start incubating. But if the nest is empty, the female of the host species is likely to desert it on finding the cowbird eggs. Yellow warblers are likely to take a different tack: they build a new nest on top of the old one, even if one or two warbler eggs are in it. Nests as high as *eight* stories have been found, the floor of each new nest covering a cowbird egg. Should the cowbird choose the nest of a robin or a catbird, her egg will promptly be punctured and/or tossed out.

The cowbird baby is off to a head start over its nestmates because cowbird eggs need only ten days' incubation. The young cowbird then gets the full attention of its foster parents. It doubles in weight the first day after hatching and begins to open its eyes on the fourth day (getting them fully open on the fifth day). On the fifth day feathers appear, and the cowbird has greatly increased in size and weight. Should any of the other eggs hatch, the cowbird nestling, larger and more vigorous, might smother or destroy them. By the time it leaves the nest, it's often twice the size of the birds feeding it.

Cowbird nestlings and fledglings probably eat a larger variety of food than any other young bird. They need only ten days as a nestling, but follow and beg from their "parents" for two or three weeks after leaving the nest. A tiny ruby-crowned kinglet has been observed feeding a fledgling cowbird; so have Grace's warblers and golden-cheeked warblers.

Whereas nestlings of non-parasitic birds crouch down and remain quiet if a stranger approaches the nest, cowbird youngsters beg from everybody and anybody. They beg from their foster parents for a few weeks, and then beg from other birds. Even after the cowbirds learn to fly, they stay around the nest for a couple of days.

However indignant this behavior makes you feel, it might be sensible to consider a statement in *Audubon Bird Guide,* by Richard H. Pough (Garden City, NY: Doubleday and Company, Inc., 1949), sponsored by the National Audubon Society: "Since the birds they most frequently impose upon continue to be about as abundant as their habitats permit, it is evident that the cowbird does not have an appreciable effect upon their population level."

In the last fifty years, though, forest fragmentation has allowed cowbirds easier access to the nests of warblers and thrushes — another case where human changes to the environment have helped an aggressive species to thrive.

In keeping with their tarnished reputation, cowbirds were once actually accused of being proponents of "free love." It has since been determined that they are monogamous for the length of the breeding season, as are most perching birds.

Cowbirds were formerly called buffalo birds because they used the beasts as beaters to stir up insects, which they then consumed. Adaptable creatures, cowbirds turned their attention to cattle as bison disappeared. Cowbirds are found most frequently around farms, pastures, and open woodlands over most of the United States. In addition to grasshoppers, leaf hoppers, and spiders, they eat grains, seeds, and berries. Except during the breeding season, they normally travel in flocks.

Most people don't try to attract cowbirds to feeders. Quite aside from our tendency to disapprove of their domestic arrangements, they consume a lot of feed. They especially like hemp, millet, scratch feed, and sunflower seeds, served on the ground, but then so do many birds people are eager to attract. Besides, cowbirds aren't fussy and will take almost anything they can get. To their credit, they're not aggressive but feed amicably with other birds. Because of their appetite and numbers, it may be you'll decide you don't especially want to attract them, but surely it's not necessary to follow the lead of earlier bird lovers and actively try to exterminate cowbirds.

Size: 6½–8¼"L
Breeding range: Newfoundland west to British Columbia, south to Mexico and Louisiana, and northern Florida

Winter range: Central New England west to California, south to Mexico, and east to Florida
Preferred habitat: Fields, woods edges, farms

Preferred nest site: Parasitic
Clutch size: 1 to 6, usually 3
Incubation period: 10 days
Nestling period: 10 days
Broods per season: 3 or 4
Preferred food: Insects, seeds, berries, grains

TANAGERS

THE family Thraupidae (formerly Tangaridae) contains singularly colorful birds that are a pleasure to see in the yard or garden. They winter in the tropics.

Left to right: Western tanager, scarlet tanager, summer tanager.

Scarlet Tanager

(Piranga olivacea)

THE scarlet tanager, a resident of deciduous woods, is easily identified by its black wings and tail. You may see it in your garden, catching insects on the wing, if you live anywhere from the Great Plains east to the Atlantic. Unfortunately, it's quite unlikely to appear at your feeder. Its food is nine-tenths insects, especially bees and wasps. However, scarlet tanagers (sometimes called black-winged redbirds, or firebirds) eat a few berries, too, like those of the euonymus.

Female scarlet tanagers build a small, flimsy, flat cup of a nest on a tree limb, preferably oak, out from the trunk and from 8 to 75 feet from the ground. It is made of twigs and rootlets, lined with grasses.

Size: 6¼"L
Breeding range: Nova Scotia west to southern Saskatchewan, south to eastern Oklahoma, and east to coast of Virginia

Winter range: South America
Preferred habitat: Deciduous and mixed woodlands
Preferred nest site: On an oak branch away from the trunk, 8 to 75 feet from the ground

Clutch size: 3 to 5, usually 4
Incubation period: 13 to 14 days
Nestling period: 14 to 16 days
Broods per season: 1
Preferred food: Insects, fruits

Western Tanager

(Piranga ludoviciana)

THE western tanager, yellow with black wings and a red head, normally ranges from the Black Hills to the Pacific, but some individuals are found in the East nearly every year during migration. They spend the winter in Central America.

Their food is insects, fruit, and berries. Western tanagers come regularly to feeders in their range, where they will sample just about anything. In the garden, they like the fruit of mountain ash and, especially, cherries. Their fondness for cherries sometimes gets them into trouble with orchardists, but their appetite for insects perhaps compensates for that transgression. The bird lover is likely to be willing to protect the cherries with netting and try to encourage this lovely bird to use the feeder instead.

The western tanager's song sounds a lot like the robin's. Little is known about its courtship behavior, but it may include food-offering. The male's plumage is much brighter during the courtship season. Western tanagers build frail, 5-inch-wide, saucer-shaped nests of grasses, bark, and twigs lined with rootlets and hair. They may be placed on branches 3 to 25 feet from the trunk of the tree and 5 to 30 feet from the ground. The chicks are thirteen to fifteen days in the nestling stage.

Size: 6¼"L
Breeding range: Southeast Alaska, south and east Mackenzie in the Northwest Territory east to the Black Hills, west Texas, west to Baja California

Winter range: Mexico to South America
Preferred habitat: Open pine and fir forests
Preferred nest site: Horizontal branch of conifer, 5 to 30 feet from the ground

Clutch size: 3 to 4
Incubation period: 13 days
Nestling period: 13 to 15 days
Broods per season: 1
Preferred food: Insects, fruits, berries

Summer Tanager   (Piranga rubra)

THE summer tanager, at home in the South, is rosy red and, unlike the cardinal, has no crest. It will readily visit feeders for soft fruits such as banana, as well as suet and peanut butter. Its appetite for bees has earned it the nickname of bee bird in some places. In addition to bees, its insect food includes wasps, cicadas, flies, weevils, and spiders. If your garden contains blackberries, mulberries, or figs, it may be attractive to the "summer redbird."

The nest of summer tanagers is similar in shape to that of scarlet tanagers but typically is made of grasses, rootlets, leaves, bark, and Spanish moss and lined with fine grasses. The female usually builds it 5 to 20 feet from the ground, although nests have been reported as high as 60 feet from the ground. It takes her about two weeks. Summer tanagers feed their nestlings for seven to ten days, the male helping.

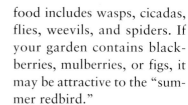

Size: 6½"L
Breeding range: Delaware west to southern California and northern Mexico, east to the Gulf states and Florida

Winter range: Mexico, Central and South America
Preferred habitat: Lowlands and foothills
Preferred nest site: Woodlands and groves

Clutch size: 3 to 4
Incubation period: 11 to 12 days
Nestling period: 7 to 10 days
Broods per season: 1
Preferred food: Insects

CARDINALS AND GROSBEAKS

THIS group's most common characteristic is the bill, which is short and stout, adapted to seed eating. These birds also eat insects, fruits, and berries, but their staple item of diet is seeds. Grosbeaks are often colorful, and females and males tend to be distinctively different in plumage. The females are much drabber in appearance than the males, probably for safety while incubating the eggs.

Incidentally, look for **evening grosbeaks** with the finches. Despite their name, they are actually more closely related to the purple and other finches than to the rosebreasted and other grosbeaks.

Cardinals are usually listed with grosbeaks because of their very heavy bills, and some people even call them cardinal grosbeaks. The female is much more subdued in appearance than the male but still easily recognized. Cardinals are usually seen in pairs, and sometimes two or more pairs visit the same feeding stations.

The cardinal may well be the most popular of all our native birds. Even the most casual observer is pleased by sighting our only crested red bird. We have a couple of pairs coming regularly to our feeders here in western Massachusetts, and friends in Iowa boast six or eight at a time. We see them most often at the honeysuckle bush during its fruiting season. Cardinals also like the hornbeam, and my Aunt Vi in Illinois has counted eleven males and five females at one time enjoying the cat-kins on her birch in January. Friends in West Virginia feed shelled peanuts to cardinals from their back porch every evening. If Kate and Bud are late in appearing, the birds fly to the kitchen window to remind them.

Left to right: Rose-breasted grosbeak, black-headed grosbeak, northern cardinal, blue grosbeak.

Northern Cardinal

(Cardinalis cardinalis)

THIS non-migratory species was formerly considered a southern bird, but it has vastly extended its range northward. Cardinals utter loud, clear whistles and beautiful songs with many variations; they are sometimes confused with the songs of the Carolina wren.

When courting, the male chases a female, fights other males, and swoops and dives. Once mated, the female cardinal builds a loose nest of twigs, leaves, weeds, bark, and grasses lined with hair, rootlets, and fine grasses, taking three to nine days. She chooses the site, usually about 8 feet from the ground but sometimes as high as 30 feet, and the male brings the materials.

Cardinals are fond of sunflower seeds, watermelon and cantaloupe seeds, scratch feed, bread, and peanut butter. Though they prefer feeding on the ground, they don't mind a hanging or stationary feeder. In the yard, cardinals will feast on almonds, white ash, American elm, and both cherry and autumn elaeagnus, as well as blackberries and honeysuckle berries and the seeds of four o'clocks. If you live in a subtropical climate, you may have discovered that they dote on camphor trees. Though cardinals are basically seed and fruit eaters, such insects as ants and beetles may account for up to a third of their food when available.

Cardinal parents feed their young for nine or ten days, eating the fecal sacs for the first four or five days and after that carrying them away. The male watches over the fledglings for about three weeks while the female nests again.

Size: 7¾"L
Breeding range: New Brunswick west to Great Plains; south to Texas, southern New Mexico, and Arizona; Mexico east to Florida

Winter range: Permanent resident in breeding range
Preferred habitat: Open woodland, parks, gardens, suburbs
Preferred nest site: Shrubs, small trees

Clutch size: 2 to 5, usually 3 or 4
Incubation period: 12 to 13 days
Nestling period: 9 to 14 days
Broods for season: 1 to 3
Preferred food: Seeds, fruits, grains, insects

256

Rose-Breasted Grosbeak (Pheucticus ludovicianus)

I ALWAYS eagerly await the arrival of the rose-breasted grosbeaks in spring; they're among my favorites because they're extraordinarily beautiful, more so even than the orioles. During courtship the males fight. They also sing and hover in the air around the female.

The female rose-breasted grosbeak is brown and streaked like a sparrow, but the male is unmistakable — black and white with a triangular red patch on his breast. They come to feeders for sunflower seeds and are attracted to barberry, beech trees, corn, cherries, and redbud, and they are reputed to have a truly admirable appetite for the Colorado potato beetle. The rose-breasted grosbeak breeds from central Canada south through the Plains states, to the Northeast and south to Georgia and Appalachia.

Both male and female rose-breasted grosbeaks sing, occasionally even (softly) while on the nest. Their songs — long, smooth, and liquid — have been compared favorably to those of the robin. The loose, saucer-shaped nests of the rose-breasted grosbeaks are made of twigs, weeds, and grasses lined with finer materials and placed 5 to 20 feet from the ground.

Size: 7¼"L
Breeding range: Nova Scotia west to northeastern British Columbia, eastern Colorado south to Oklahoma and east through Ohio and New Jersey; mountains to northern Georgia

Winter range: Mexico to South America
Preferred habitat: Edges of moist second-growth woods; suburbs
Preferred nest site: Fork of deciduous tree, from 5 to 25 feet from ground

Clutch size: 3 to 6, usually 4
Incubation period: 12 to 14 days
Nestling period: 9 to 12 days
Broods per season: 1, sometimes 2
Preferred food: Insects, seeds, fruits

Black-Headed Grosbeak *(Pheucticus melanocephalus)*

THE western counterpart of the rose-breasted grosbeak is the black-headed grosbeak, also known as the western grosbeak, which is remarkably similar in song and quiet temperament but not as brilliant in plumage. Its fondness for early fruit, peas, and beans is balanced by its consumption of scale insects (about a quarter of its diet), beetles, canker worms, and snails. Although its diet is more than two-thirds animal in origin, it will visit feeders for bread and sunflower seeds.

The male arrives in the breeding territory before the female and sings to stake out his claim. The black-headed grosbeak's nest is bulky, made of interwoven twigs, plants, and rootlets, built in a large shrub or small tree near a stream. Both parents incubate the three to four eggs.

Size: 7¼"L
Breeding range: Southern British Columbia east to western North Dakota south to northeast Kansas, south through western United States to southern Mexico

Winter range: Mexico
Preferred habitat: Mixed woods, piñons, orchards, parks
Preferred nest site: Tree or bush
Clutch size: 3 to 4
Incubation period: Probably 12 days

Nestling period: Usually 12 days
Broods per season: 1
Preferred food: Insects, fruits

Blue Grosbeak

(Guiraca caerulea)

THE blue grosbeak tends more to the southern section of the United States. Like its western cousin, it consumes some insects we consider harmful. Blue grosbeaks are sometimes cursed for invading grainfields for sorghum, oats, and corn. Blackberries and sunflowers are apt to attract them, but gardeners are often willing to forgive them their occasional lapses because they're also fond of crabgrass seed.

Size: 6¼"L
Breeding range: North-central California east to New Jersey, south to Florida, west through north-central Mexico to north-central Baja California

Winter range: South central Arizona through Mexico to Panama
Preferred habitat: Brush, thickets, farms
Preferred nest site: Bush or low tree

Clutch size: 3 to 5
Incubation period: 11 days
Nestling period: 13 days
Broods per season: 2, at least in the South
Preferred food: Insects, grains

BUNTINGS

Left to right: Indigo bunting, lazuli bunting, painted bunting, snow bunting.

WOOD WARBLERS, SPARROWS, BLACKBIRDS, AND TANAGERS   *Emberizidae*

Lazuli Bunting   *(Passerina amoena)*

THE lazuli bunting is the western counterpart of the indigo bunting, occurring from the Great Plains to the Pacific Coast, except in Texas. Their ranges overlap, and the two hybridize so regularly that some ornithologists believe they should be treated as one species. Where the range doesn't overlap, however, the plumage is distinctly different, with the lazuli bunting having white wing bars and a chestnut breast. Some birders believe them to be wary of humans, and they are rare visitors to feeding stations. If they are common in your area, you might try attracting them with the foods that indigo buntings enjoy.

Size: 4½"L
Breeding range: Great Plains and southern Saskatchewan west to British Columbia and south to northwestern Baja California

Winter range: Southeastern Arizona, Mexico
Preferred habitat: Dry brushy areas
Preferred nest site: Low brush near water

Clutch size: 3 to 4
Incubation period: 12 days
Nestling period: 10 to 12 days
Broods per season: 2
Preferred food: Seeds, insects

Indigo Bunting (*Passerina cyanea*)

YOUR first sighting of an indigo bunting is apt to leave you gasping with disbelief: they're astonishingly brilliant. It's the only native bird that is dark blue all over. Because they like brush, young second growth, orchards, and pastures, they're much more common now than they were during colonial times. They seem to enjoy the heat of the day in July and August, singing persistently when other birds retire to trees and thickets. They eat blackberries, corn, and dandelion seeds as well as beetles and caterpillars. Sometimes they come to feeders for wheat or nutmeats, and they may be attracted to a birdbath.

The indigo bunting breeds throughout the United States from the Great Plains east, excluding southern Florida and southern Texas. As many as fifty-two pairs of indigo buntings have been reported on one hundred acres of favorable habitat in Maryland. Defense consists of the male performing its finchlike song from a perch; it ceases with the August molt.

Once paired, indigo buntings nest in both maples and briar patches, where they make good cup-shaped nests of dried grasses, twigs, leaves, and weeds, lined with fine grasses, hair, feathers, and rootlets on a base of leaves or snakeskin. The nests are 3¼ to 4½ inches in diameter and 2½ to 3 inches high and are usually hidden in underbrush or a thicket, 4 to 12 feet from the ground. The birds may use the same nest for as long as five years, repairing it as needed.

Size: 5½"L
Breeding range: Maine west to southern Manitoba, southwest to southeastern California and east to northern Florida; absent from west Texas, Oklahoma panhandle

Winter range: Central Mexico south to Panama, Cuba, Jamaica, and the Bahamas; south Florida
Preferred habitat: Roadsides, brush, edges of woods
Preferred nest site: Underbrush, thickets

Clutch size: 2 to 6, usually 3 or 4
Incubation period: 12 to 13 days
Nestling period: 10 to 13 days
Broods per season: Often 2
Preferred food: Insects, seeds

Painted Bunting (*Passerina ciris*)

WITH luck and the right location, you might glimpse the gorgeous painted bunting when it struts and spreads its wings and tail. The most distinctive of the buntings, the male painted bunting has a purple head, green back, and red breast. It was once a popular cage bird because of its spectacular plumage. Pugnacious about territory, it was easy to trap for sale, using a dummy or stuffed bird as a lure. The female is green, the only native finch that can claim that distinction. Painted buntings are found from North Carolina west to Kansas and thence south.

They feed on weed and grass seeds, pine seeds, sunflower seeds, and rice, as well as flies, wasps, and boll weevils. Insects account for about a quarter of their diet. At feeders they eat mixed seeds, peanut hearts, cracked corn, and bread crumbs. Like the indigo bunting, the painted bunting has a finchlike song.

Painted buntings nest in elms, oaks, and blackberry bushes. They choose a site in thick foliage, usually 3 to 6 feet but sometimes as high as 20 feet from the ground. The cup-shaped nest is constructed of twigs, leaves, rootlets, bark, weeds, and grasses bound with caterpillar silk and lined with fine grasses, rootlets, and horsehair. It is similar to an indigo bunting's nest but neater.

Although they're considered shy, painted buntings are so belligerent toward other males that sometimes their fights are fatal.

Size: 4½"L
Breeding range: Kansas southeast to Florida panhandle, south along Gulf coast to northeastern Mexico, central New Mexico; South Carolina to northeastern Florida

Winter range: Southern Florida; Mexico to Panama and Cuba
Preferred habitat: Thickets, hedgerows, gardens
Preferred nest site: Crotch of bush
Clutch size: 3 to 4

Incubation period: 10 to 12 days
Nestling period: 12 to 14 days
Broods per season: 2 to 4
Preferred food: Seeds, insects

Snow Bunting

THE snow bunting is nowhere common at feeders, and if we see it at all in the North, it will be in winter. It is our whitest land bird. The male has black feathers on back, wings, and tail; the female is similar, with rusty instead of black markings. At seven inches, it's larger than the other buntings. It feeds on amaranth, corn, wheat, and barley, supplemented by beetles and caterpillars in season. Encountering one in your garden can be considered a coup. Scatter some cracked corn or oats to please these rare visitors.

Size: 8"L
Breeding range: Northern Alaska to Greenland, south to north Quebec
Winter range: Southeast Alaska east to central Quebec, south to northern United States

Preferred habitat: Open, barren, rocky land or grasslands
Preferred nest site: Crevice among rocks or grasses
Clutch size: 4 to 6

Incubation period: 14 days
Broods per season: 1
Preferred food: Weed seeds, insects

Dickcissel (*Spiza americana*)

THE dickcissel causes a great commotion in the birder's world because it is irregular in its habits. At one time, for example, it was quite common along the Atlantic seaboard, but then it disappeared entirely in that area. (It may be making a comeback now.) Though it's fairly abundant in the Midwest, its appearance even there is sporadic. It isn't known whether these population fluctuations are a result of diseases or migration disasters.

The dickcissel is an attractive bird of sparrow size with streaked brown back, yellow breast, and black bib. It's considered beneficial because it has more interest in insect food than most finches, consuming crickets, ants, flies, and the like to account for about four-fifths of its diet. It will visit feeders for peanut butter, millet, oats, wheat, and corn.

Typically the dickcissel nests in elm, mulberry, or Osage orange trees. The female builds the large and bulky nest, made of weeds and grasses and lined with softer materials, about 15 feet from the ground. Dickcissel chicks are in the nestling stage for ten to twelve days.

Dickcissels are early migrators, packing up in August immediately after the nesting season is over and heading for South America in flocks of two hundred or three hundred birds. The dickcissel (of course) says *dick-dick-cissel-cissel-cissel*.

Size: 6¼"L
Breeding range: Southern Michigan west to southeastern Saskatchewan and south to Texas, east to south central Pennsylvania

Winter range: South America
Preferred habitat: Open spaces, fields, prairies
Preferred nest site: On or close to the ground, concealed by plants

Clutch size: 3 to 5, usually 4
Incubation period: 10 to 13 days
Nestling period: 10 to 12 days
Broods per season: 2
Preferred food: Seeds, grains, insects

Green-Tailed Towhee

(Pipilo chlorura)

THE green-tailed towhee, found from the Great Plains west, is smaller than its eastern cousin. The edges and underparts of the wing are yellow, the cap rufous (reddish), the throat white.

Towhees eat spiders and snails, ants, beetles, flies, wasps, grasshoppers, and caterpillars as part of their diet. To lure these ground feeders to your feeding area, use peanut chips, nutmeats, cracked corn, and sunflower and watermelon seeds. They dote on the crumbs of bread and crackers, and they love to bathe in birdbaths. They'll gather in the sumac in your "natural" area and may be found dining on raspberries, blackberries, blueberries, and strawberries. You'll know their location because they make a terrific racket, scratching like domestic chickens.

In courtship, both males and females rapidly open and close their wings and tails in display. The females usually build, on or near the ground, bulky but firm nests of leaves, weeds, twigs, ferns, moss rootlets, bark, and grasses, lined with fine grasses, pine needles, and hair. Building takes about five days. The nestlings stay in the nest ten to twelve days, and the male helps feed the young.

The female is apt to perform the injured-wing routine if you surprise her on or near the nest. If you're simply in the vicinity, she'll leave the nest and, when she judges herself to be a safe distance away, attempt some activity to draw attention to herself.

Cowbirds frequently choose towhee nests for parasitizing. If more than two or three cowbird eggs are deposited, the towhee deserts the nest, apparently realizing she's been had.

Size: 6¼" L
Breeding range: Great Plains west, western Montana south to southern California, Arizona, and New Mexico

Winter range: Southeastern California, Arizona, and New Mexico south to central Mexico
Preferred habitat: Brushy hillsides
Preferred nest site: Underbrush
Clutch size: 3 to 4

Incubation period: 12 to 13 days
Nestling period: 10 to 12 days
Broods per season: 1 to 3
Preferred food: Seeds, wild fruits, insects

Rufous-Sided Towhee

(Pipilo erythrophthalmus)

RUFOUS-sided towhees are larger than green-tailed towhees. Males have white breasts, black heads, backs, and wings, and robin-colored sides. White spots in the tail are conspicuous in flight, and there are white markings on the wing. In the female, black is replaced with brown. These birds can be found, at one season or another, from coast to coast. A slightly smaller western variation, with considerably more white on the wings and some on the back, is called the spotted towhee or the Oregon towhee.

If you're out in the woods in springtime and think you hear an army of moose approaching, it may only be a rufous-sided towhee chasing a female he fancies. The amount of racket towhees can make as they go about their activities is impressive.

The handsome, robin-sized towhee is familiarly known as "chewink" — to say nothing of "swamp robin," "ground robin," "marsh robin," "bullfinch," "bushbird," and "turkey sparrow." Its more common names come from its call, transliterated both as *towhee* and *chewink*, which shows why I distrust written descriptions of bird songs. The same call has been written as *joree-ziee, t'wee, shrink,* and *cherwink*. Its song is variously written as *"Drink your tea, see towhee"* and *"Sweet bird sing"* and *"Cheet, cheet, cheer"* and *"Cheap, cheap cheese."* Some of these variations may be a result of differences in songs and calls in different parts of the country, but even so, they strike me as less than helpful.

Several different species and subspecies of towhees have been described. Ornithologists now have decided that eastern and spotted towhees are subspecies of rufous-sided towhees. Green-tailed, canyon, California, and Abert's are distinct species.

Size: 7¼"L
Breeding range: Southern Maine west to southern British Columbia, south to Mexico, east to Florida
Winter range: New Jersey west to Texas, north to southern British Columbia; retreats from mountains

Preferred habitat: Clearings in or edges of woods, brush, pastures, shrubs
Preferred nest site: On ground in brush, low in shrub
Clutch size: 3 to 6, usually 3 or 4

Incubation period: 12 to 13 days
Nestling period: 10 to 12 days
Broods per season: 1 to 3, usually 2
Preferred food: Insects, seeds, fruits, mast

Dark-Eyed Junco   (*Junco hyemalis*)

ABOUT the general size and shape of a sparrow is that gregarious bird known as the dark-eyed junco, or snowbird. Juncos are quite variable, and many subspecies have been named, but they always have white outer tail feathers and a white belly. The rest of their plumage is gray or black, without streaks, but sometimes rufous or pinkish on the sides. They hybridize very occasionally with white-throated sparrows.

Depending on the season, juncos can be found all over the United States, but they breed mostly in Alaska, Canada, and New England, south through mountainous areas to Georgia. They prefer feeding on the ground, their favorite feeder foods being cracked corn, peanuts, millet, suet, sunflower seeds, and wheat. In the summertime juncos thrive on ants, spiders, caterpillars, and grasshoppers, in addition to weed and grass seeds and a few wild berries.

Except during the breeding season, juncos are usually found in flocks averaging sixteen in number. They associate not only with other juncos but also with towhees, yellowrumped warblers, and fox and white-throated sparrows during the winter.

The dark-eyed junco makes a bulky nest on sloping ground under weeds and grasses. The female constructs it of grasses, rootlets, moss, and bark shreds lined with grasses and cow or deer hair. The young spend twelve days in the nest.

Snow is a problem to these hardy little birds when it covers their natural food. At such times, flocks are likely to wander extensively in search of bare ground. They've been sighted on coastal beaches during periods of heavy snow cover.

Wherever juncos are found, they're welcome visitors, attractive and well mannered.

Size: 5¼"L
Breeding range: Labrador west to Alaska, south to northeastern New Mexico; northern Minnesota to Pennsylvania and south in mountains to northern Georgia

Winter range: Retreats from northernmost part of breeding range; winters throughout United States, except south Florida
Preferred habitat: Coniferous and mixed woods
Preferred nest site: On ground

Clutch size: 3 to 6, usually 4 or 5
Incubation period: 12 to 13 days
Nestling period: 9 to 12 days
Broods per season: 1 or 2
Preferred food: Insects, wild fruits, seeds

SPARROWS

SPARROWS offer a real problem in identification to the novice. There are so many different kinds with such subtle differences. We'll just briefly mention those that rarely or never come to feeders: Bachman's, Cape Sable, clay-colored, grasshopper, Henslow's, Ipswich, lark, Le Conte's, Lincoln's, savannah, seaside, sharptailed, swamp, and vesper.

Because sparrows nest and forage on and near the ground, they're subject to certain dangers usually avoided by more arboreal birds. For example, they are among the prey of bullfrogs (which, in turn, may be the prey of crows, grackles, and owls). The young develop very rapidly and often leave the nest before they're able to fly. The male takes over feeding chores while the female gets busy on another brood.

Both the chipping sparrow and the field sparrow have an interesting characteristic as nestlings, perhaps a throwback to their reptilian origins: They are cold-blooded when hatched. Both sparrows become warm-blooded by the age of seven days.

Clockwise from upper left: Song sparrow, white-crowned sparrow, white-throated sparrow, tree sparrow, fox sparrow, Harris's sparrow, chipping sparrow,; center, field sparrow.

Harris's Sparrow   *(Zonotrichia querula)*

COMMON to the Great Plains, Harris's sparrow is our largest sparrow, distinctive in appearance with black cap, face, and bib. In nature its food is mostly weed and grass seeds, although it likes some fruits such as cranberries. It will take corn, sunflower seeds, hemp, or wheat on the ground or at window feeders.

Size: 7"L
Breeding range: Northwest Territories, northern Saskatchewan, Manitoba, northernmost Ontario

Winter range: Eastern Colorado to southwest Iowa, south to central Texas
Preferred habitat: Dwarf timber, thickets

Preferred nest site: On ground
Clutch size: 3 to 5
Incubation period: 12 to 14 days
Nestlings period: 13 to 15 days
Broods per season: 1
Preferred food: Seeds

Tree Sparrow   *(Spizella arborea)*

THE tree sparrow has a reddish cap but is distinguished from the chipping sparrow by the black spot on its breast. Tree sparrows are found over most of the country but are more abundant along the Pacific Coast. During the winter they're often seen at feeders, where they're relatively tame. They will peck at suet, and they also like cracked corn, hemp, peanut butter, and pumpkin seeds. Their natural food consists of weed and grass seeds and, in summer, a few insects. The tree sparrow says *whee-hee-ho-hee* and calls *tseet*.

Tree sparrows nest in thickets and make bulky cups of grasses, rootlets, weeds, bark, and feathers, lined with animal hair or fur.

Size: 5¼"L
Breeding range: Along edge of tundra from Labrador to Alaska
Winter range: Maritime provinces west to southeastern British Columbia, south to Nevada, northern New Mexico to Virginia

Preferred habitat: Open country
Preferred nest site: Thickets
Clutch size: 4 to 5
Incubation period: 11 to 14 days

Nestling period: 12 to 14 days
Broods per season: 1
Preferred food: Seeds

Chipping Sparrow *(Spizella passerina)*

THE chipping sparrow is sometimes called the "social sparrow" because it's tame in the manner of robins. It adapted readily to our growing population but was for some time threatened by the imported house sparrows. With the decline in numbers of house sparrows, the chipping sparrow rapidly expanded its population. Although found in the West, it is most common in the eastern United States. Chipping sparrows feed on insects, plant seeds, and occasionally berries. At the feeder, they'll take cracked corn, millet, oats, peanut butter, and suet. You'll recognize the slender bird by its gray breast, red-dish cap, forked tail, and white eye stripe.

Chipping sparrows nest in trees, vines, or shrubbery, usually 1 to 25 (but up to 40) feet from the ground. The nest is made of fine grasses, weeds, twigs, and rootlets, lined with fine grasses and cow, horse, or deer hair. Both parents build, in three or four days, a nest 4⅜ inches in diameter and 2¼ inches high. The nest is frequently victimized by cowbirds, for which the chipping sparrow is their favorite host. Chipping sparrows stay in the nest nine or ten days.

Size: 4¾"L
Breeding range: Western Newfoundland west to Alaska, south to Central America, east to Gulf states and Georgia
Winter range: Southern Maryland west to Texas and southern California and south

Preferred habitat: Pine woods, edges of woods, farms, gardens
Preferred nest site: Vines, shrubbery, tree
Clutch size: 2 to 5, usually 3 or 4

Incubation period: 11 to 14 days
Nestling period: 7 to 10 days
Broods per season: 1 to 3, usually 2
Preferred food: Insects, seeds

271

Field Sparrow   *(Spizella pusilla)*

THE field sparrow is primarily an eastern sparrow. It's similar in appearance to the tree sparrow but without the black spot on his breast. Field

sparrows commonly flock with chipping sparrows and like essentially the same feeder tidbits: cracked corn, corn bread, proso millet, peanut butter, and suet. Their natural food is primarily the seeds of grass and weeds; in summer they augment their diet with many insects, such as tent caterpillars, beetles, ants, and flies.

The field sparrow nests from ground level to a height of 10 feet and makes its nest of grasses, weeds, and rootlets lined with grasses and hair in three to five days. Field sparrow young stay in the nest nine or ten days.

Size: 5"L
Breeding range: Central New England to eastern Montana, south to Texas panhandle, south to the Gulf states and east to the Atlantic; west-central Montana
Winter range: Southern New England west to southeastern

Colorado, south to Texas and Mexico, east to Florida
Preferred habitat: Second growth, brush, pastures
Preferred nest site: On ground or in shrubs
Clutch size: 2 to 5, usually 3 or 4

Incubation period: 11 to 13 days
Nestling period: 7 to 10 days
Broods per season: 1 to 3
Preferred food: Insects, seeds

White-Crowned Sparrow   (*Zonotrichia leucophrys*)

THE white-crowned sparrow is found more often in the West. Its very rarity makes it highly esteemed by eastern birders. In summer its diet is largely insects, but at other seasons it becomes a weed-and-grass-seed eater, supplemented with waste grain and wild fruit. At feeders it likes cracked corn, sunflower seeds, and walnut meats.

White-crowned sparrows prefer open, hilly, brushy country, where they nest on the ground. The nests are made of small twigs, grasses, leaves, rootlets, and moss and are lined with fur, hair, fine grasses, bark, and rootlets.

The female takes over incubation duties.

White-crowned sparrows migrate in flocks. As they gather before departure they seem to whisper to each other, especially during rainy weather.

Size: 5¾"L
Breeding range: Breeds near the tree line in Alaska and Canada, south along the coast and in the high mountains to southern California, central Arizona and northern New Mexico

Winter range: Southwestern British Columbia south to central Mexico, Ohio Valley and southern New Jersey to Louisiana
Preferred habitat: Brush, thickets; in winter, open scrub, roadsides, gardens

Preferred nest site: On ground or near ground in a bush
Clutch size: 3 to 5; 6 in the Arctic
Incubation period: 12 to 14 days; 15 to 16 days in Alaska
Nestling period: 15 to 16 days
Broods per season: 1
Preferred food: Insects, seeds

White-Throated Sparrow   *(Zonotrichia albicollis)*

THE white-throated sparrow has a beautiful song consisting of whistles, opening usually with a couple of clear notes and followed by three quavering notes of different pitch. Some people insist the bird is saying *Old Sam Peabody, Peabody, Peabody* — but, then, people say remarkable things sometimes.

The white-throated sparrow is more apt to be found in the eastern portion of the country, where it thrives on insects, berries, and weed seeds. At the feeder it prefers peanut hearts and cracked corn, and it is adept at the art of thievery.

Female white-throated sparrows build cups of grasses, twigs, leaves, and moss, lined with grasses, rootlets, and hair. White-throated sparrows stay in the nest twelve to fourteen days and are fed by both parents.

Size: 5¾"L
Breeding range: Labrador west to southern Yukon, south to Wisconsin, Pennsylvania, New England

Winter range: Ohio valley and central New England south to Mexico, the Gulf states, and Florida; coastal California
Preferred habitat: Woodland undergrowth, brush
Preferred nest site: On ground

Clutch size: 3 to 5, usually 4
Incubation period: 11 to 14 days
Nestling period: 12 to 14 days
Broods per season: Often 2
Preferred food: Insects, seeds, wild fruits

274

Fox Sparrow *(Passerella iliaca)*

THE fox sparrow, so called because of its reddish hue, is a large sparrow and perhaps the easiest one to recognize. He's found all over the continental United States and has a larger number of subspecies than any other sparrow except the song sparrow. Fox sparrows eat weed seeds; birch and alder; wild fruit; the berries of holly, euonymus, and cedar, and insects. At the feeder they'll accept millet, cracked corn, and suet. They're more likely to come in times of heavy snow cover than in open weather.

Fox sparrows nest on the ground and up to 12 feet; their cup-shaped nests are made of grasses, moss, weeds, bark, and rootlets and lined with feathers, grasses, and rootlets. Fox sparrows have an eleven- to thirteen-day nestling period.

Size: 6¼"L
Breeding range: Labrador west to Alaska to the tree line, south through the high mountains of California, Nevada, Utah, Colorado

Winter range: Coastal Massachusetts west to southern California and north to southwestern British Columbia
Preferred habitat: Stunted boreal woodlands and undergrowth; in winter, undergrowth, parks, gardens

Preferred nest site: In a bush or on the ground
Clutch size: 3 to 5
Incubation period: 12 to 14 days
Nestling period: 11 to 13 days
Broods per season: 2
Preferred food: Insects, seeds, fruits

275

Song Sparrow

(Melospiza melodia)

THE song sparrow puts all his relatives to shame with his astonishing music. He is cheerful and persistent, with great variation. At least two dozen songs comprise his repertoire, all of them beginning with three to six emphasized notes. A common beginning is the first four notes of Beethoven's Fifth Symphony. Although each song sparrow has songs peculiar to itself, all use bits and pieces of their neighbors' songs. They sing to each other, and the songs sound like human songs that they never get around to finishing.

The song sparrow is found all over the country, according to local environmental conditions (at least twenty-five subspecies have been described). Song sparrows tend to be visible; they're aggressive and chase other males as well as females — one assumes with different intentions.

Most song sparrows migrate south of their nesting areas in winter, but not very far and not for long. They're largely terrestrial, and their colorings reflect their habits. They're identified by heavily streaked underparts forming a dark blotch in the middle of the breast. Although song sparrows are not as adaptable to close contact with people as chipping sparrows, their sweet and varied songs have made them well loved. They'll come to feeders for hemp, millet, and peanut butter. Occasionally they sample suet and cracked corn. They prefer feeding on the ground on fallen or scattered seed but will learn to use hanging or stationary feeders.

By wintertime, about 90 percent of the song sparrow's diet is vegetable, consisting largely of seeds, especially weed seeds.

Size: 5½"L
Breeding range: Newfoundland to the Aleutians, south to Mexico and east to the mountains of northern Georgia
Winter range: Central Maine west to Washington, north to the Aleutians, south to central Mexico, east to the Gulf states and Florida
Preferred habitat: Thickets, shrubbery
Preferred nest site: On the ground or in a shrub
Clutch size: 3 to 6
Incubation period: 12 to 13 days
Nestling period: 10 to 14 days
Broods per season: 2, sometimes 3
Preferred food: Insects, seeds, fruits

Song sparrows defend their half- to one-and-a-half acre territories by singing from perches and puffing out their feathers; the defense ceases during the late summer. Blackberries, raspberries, and elderberries will provide nesting sites as well as food for them. Song sparrows nest on or near the ground, in shrubs or high weeds. The female builds the cup-shaped nest of grasses, bark, and leaves gathered by both parents. The inside diameter of the cup, which takes about two days to build, is 2¼ inches; the outside diameter is a relatively enormous 5 to 9 inches, and it's 4½ inches high. It is lined with grasses, hair, and rootlets.

Song sparrow parents brood their young for the first five days of their eight- to ten-day nestling period, feeding them insects. The fledgling stage lasts nearly three weeks, and for the first week of which the young can't fly well. Both parents feed them at this stage, too, unless the female starts a new brood. In that case the male works alone. Nestlings are fed mostly insects or larvae found on or near the ground.

FINCHES

WITH their short, stout bills and sociable natures, the seed-loving finches are frequent visitors to feeders. For those interested in classification, the finch family, or Fringillidae, is no longer the largest group of birds, having included sparrows, buntings, and all of the grosbeaks. Scientists have now assigned those birds, including most of the grosbeaks, to the Emberizidae family.

One exception is the splendid **evening grosbeak,** which despite its name is genetically closer to the **goldfinch** than to other grosbeaks. At eight inches in length it is a good three inches longer than the sparrowlike goldfinch, although their coloring is similar.

Clockwise from upper left: Pine siskin, purple finch, red crossbill, evening grosbeak, pine grosbeak, common redpoll, American goldfinch, house finch.

Fringillidae

Evening Grosbeak

(Coccothraustes vespertina)

LIKE the cardinal, the evening grosbeak has extended its range, and even more dramatically. This boldly colored yellow bird with black-and-white wings was once rare except in the Northwest. Since the turn of the century it has become increasingly widely distributed and now winters regularly as far south as the Washington-Baltimore area on a line extending westward through Ohio. Some individuals nest as far south as New Jersey. Grosbeaks have been reported in all of the contiguous forty-eight states including northern Florida and Louisiana.

With their powerful bills, evening grosbeaks are able to extract the kernels of large seeds. Some people consider them pests because of their gluttony, but it's probably more common for those who stock feeders to report their appetites with a kind of awe. Though they're willing to eat most seeds, they are especially fond of sunflower seeds, peanuts, and safflower seeds. Away from the feeder, they eat both buds and seeds of box elders, which are widely planted from the Plains States to the East. Cottonwoods and catalpas also attract them, as well as white ash, elm, beech, maple, dogwood, sumac, and honeysuckle. Their diet is nearly all of vegetable origin, but in the summer they dine on some insects and feed considerable numbers to nestlings. They're gregarious birds and customarily live in good-size flocks.

The evening grosbeak utters short loud warbles and whistles during courtship. These birds prefer conifers as nesting sites, 15 to 60 feet from the ground. Even when nesting, they aren't strongly territorial and are relatively social, although usually there's but one nest to a tree. Their frail, elliptical cups are made of twigs and moss, 5 to 5⅞ inches in diameter and 5 inches high. The lining is made of grasses or rootlets. Evening grosbeaks are in the nestling stage for fourteen to sixteen days.

Size: 7¼"L
Breeding range: Nova Scotia west along northern Great Lakes to British Columbia, south to California and mountains of western Mexico

Winter range: Central New England west to southern Alberta, south to eastern New Mexico, east to North Carolina
Preferred habitat: Coniferous forests
Preferred nest site: Conifer

Clutch size: 2 to 5, usually 3 or 4
Incubation period: 12 to 14 days
Nestling period: 14 to 16 days
Broods per season: 1, possibly 2
Preferred food: Buds, fruits, seeds, insects

Purple Finch

PURPLE finches and house finches are closely related and similar in appearance. The former are more widely distributed and tend to prefer forests, while house finches like arid, open country. I've heard it said that purple finches haven't established the friendly relationship with humans that house finches have, a statement that astonishes me. Our feeders are filled with purple finches all summer long, with somewhat fewer in winter. They like squash and pumpkin seeds, bread, scratch feed, millet, hemp, nutmeats, and sunflower seeds.

Purple finches will use trays, hoppers, or swinging feeders with equal enthusiasm. If they feel I'm tardy in refilling a feeder, they tell me about it. They allow me to approach

very close before they retreat to a nearby tree or bush. Sometimes I see large numbers of them, often in the company of goldfinches and house sparrows, foraging on the ground near the feeders. Among the landscape plants they feed on are ash, American elm, birch, beech, box elder, balsam fir, spruce, Amur honeysuckle, privet, butternut, cotoneaster, and smoke tree.

Purple finches nest 5 to 60

feet from the ground on a horizontal branch of a conifer. The female gathers most of the material, though the male brings a bit, too. The female also does all the building of the neat cup made of twigs, weeds, grasses, and bark, lined with fine grasses or hair. The song of the purple finch is rapid and jumbled, similar to that of the house finch in its cheerful and sprightly warbling.

Size: 4½"L
Breeding range: Nova Scotia west to southern Yukon, south through Pacific states; in East, to mountains of West Virginia, west to southern Manitoba

Winter range: Southern Canadian Maritimes west to southern Manitoba, south to Texas and northern Florida
Preferred habitat: Edges of woods, coniferous forests, shade trees
Preferred nest site: Conifer

Clutch size: 3 to 6, usually 4 or 5
Incubation period: 13 days
Nestling period: 12 to 14 days
Broods per season: 1, sometimes 2
Preferred food: Seeds, buds, fruits, insects

House Finch

(*Carpodacus mexicanus*)

HOUSE finches are a western bird. They were illegally introduced into New York City about forty years ago as cage birds. When the dealers discovered their legal peril, they released the finches, which now are all over the eastern half of the United States as well as their native haunts of Mexico, the southwestern United States, and California north to British Columbia. They're also found in Hawaii, though no one is certain how they got there. They are extremely abundant in their normal range and, as their name suggests, have readily adapted to human settlements.

Interestingly, there are regional differences in the birds. In humid areas their plumage is darker; in the desert it's very bright. The eastern birds, for reasons so far unexplained, have shorter tails, wings, legs,

and toes and bigger bills than their western relatives. They're also said to exhibit less variety in song. House finches aren't fussy about nesting sites and sometimes use the abandoned homes of other birds.

The bills of house finches attest to their food preference — seeds — but they've acquired a bad reputation among fruit growers because they frequent orchards, where their numbers make them a problem. One writer reported that house finches that had gorged on rotting peaches in her garden became quite silly in behavior. At feeders, they hold their own even against

house sparrows. They prefer sunflower seeds above other feeder fare but are willing to make do with hemp, millet, cracked corn, fruits, bread, peanuts, and suet.

House finches will nest just about anywhere, including on large buildings or in the abandoned homes of other birds. Materials used are twigs, grasses, and debris of various kinds. When courting, the house finch circles the female with short, strutting hops, droops its wings, and raises the feathers of crown and crest. Both parents feed the nestlings insects and berries for twelve to fourteen days.

Size: 5¼"L
Breeding range: British Columbia south to central Texas and Mexico; introduced: central New England, New York, Michigan south to Tennessee and east to Georgia
Winter range: Permanent resident in breeding range;

also winters west Tennessee to Gulf, east to northern Florida
Preferred habitat: Open woods, inhabited areas
Preferred nest site: Practically anywhere
Clutch size: 2 to 6, usually 4 or 5

Incubation period: 12 to 16 days
Nestling period: 12 to 14 days usually; ranges from 11 to 19 days
Broods per season: 1 to 3
Preferred food: Seeds, fruits, insects

Pine Grosbeak

(Pinicola enucleator)

THE pine grosbeak is reddish with white wingbars. Altogether it reminds one of a very large purple finch. It's fairly widely distributed over the northern part of the continent and extends south to Mexico in western coniferous forests. Almost all of its diet is vegetable. Although it rarely visits feeders, it will be happy to dine in your garden if you can supply cone-bearing evergreens, apples, sumac, maples, willows, ash, blackberries, or grapes.

The pine grosbeak builds its nest of moss and small twigs, lined with hair, in an evergreen tree. The female takes charge of incubating the three or four greenish, purple-blotched eggs.

Even though a flock of pine grosbeaks can strip a tree or bush of fruit, these birds are rare enough in the southern part of their range that they are welcome anyway.

Size: 9¾"L
Breeding range: Northwestern Canada south to the northern border of the United States

Winter range: Northern half of United States, south to Kansas and Kentucky
Preferred habitat: Coniferous forests
Preferred nest site: Conifer

Clutch size: 3 or 4
Incubation period: 14 days
Nestling period: 14 to 16 days
Broods per season: 1
Preferred food: Evergreen seeds, buds

Common Redpoll

(*Carduelis flammea*)

THE common redpoll isn't quite as widely distributed as the American goldfinch, its range being more northerly. Its diet is composed entirely of plant food. Birch and alder are its favorite seeds, but it also likes dandelion, ragweed, and catalpa seeds. At feeders it likes rolled oats, hemp, millet, and sunflower seeds.

This small bird can be mistaken for a sparrow until you spot its bright red cap. It nests in thickets in the far North, building with twigs and plants and lining the cup with feathers, down, or hair. The female does the incubating.

Size: 5"L
Breeding range: Arctic; Newfoundland west to Alaska, east to coastal Greenland
Winter range: Wanders irregularly, south to Virginia, west through Nebraska, northern Oregon

Preferred habitat: Tundra, clearings in woods, swamps, fields
Preferred nest site: Low bush or tree
Clutch size: 5 to 6

Incubation period: 14 to 15 days
Nestling period: 12 to 14 days
Broods per season: 1 or 2
Preferred food: Seeds

Pine Siskin

(Carduelis pinus)

PINE siskins, the most common winter finch, are considered erratic nomads. Found all over the country, they'll visit your feeder if they're in the neighborhood. They show aggressive behavior at feeders and are not above stealing seeds from other birds. If you serve millet, hemp, nutmeats, or sunflower seeds, be prepared for droves of these finches. Apart from feeder treats, they like birch seeds and buds, alder, and Scotch thistle.

The pine siskin's song is like those of the goldfinch and the canary, though a trifle less musical. The nesting pair build a shallow saucer concealed in an evergreen 8 to 30 feet from the ground. The nest is made of twigs, bark, and moss, densely lined with fur, hair, plant down, and moss.

Size: 4½"L
Breeding range: Southern Labrador west to southern Alaska, south to northern Indiana and south along coast to central California; California south to western Mexico

Winter range: Breeding range south to Mexico and east to Florida; withdraws from northernmost portion of breeding range
Preferred habitat: Conifers, thickets
Preferred nest site: Conifer

Clutch size: 2 to 6, usually 3 or 4
Incubation period: 1 to 14 days
Nestling period: 12 to 15 days
Broods per season: 1
Preferred food: Insects, buds, seeds

American Goldfinch

(Carduelis tristis)

PERHAPS the best known of the numerous finches is the American goldfinch, or "wild canary," found in numbers all over the continental United States. Although there's a discernible shift of population in winter, goldfinches remain within their normal range all year. Their diet is nearly all seeds. At the feeder they'll eat thistle, hemp, millet, sunflower seeds, and nutmeats, but they're just as happy with the seeds of garden flowers and weeds. Among their favorites are wild thistle, catnip, coreopsis, dandelion, and cosmos. They also like goldenrod, hornbeam, birch, larch, and Chinese and Siberian elms. They won't nest where there isn't a good supply of wild thistle. Not only do they feed on its seeds, but they also use the down of the seedpod to line their nests.

There are good reasons why most of us feel drawn to goldfinches. We probably associate them with pet birds, as witness their common name and their loud, canarylike song. They are conspicuous because of their easily recognizable, attractive plumage and because they're usually found in numbers. When feeding, they seem almost ridiculously tame. They visit our flowers as soon as the seeds begin to ripen and rise in clouds when we walk past; but they don't go far, nor do they seem alarmed. I've often felt some fear for them feeding so close to the ground because of our extremely large local cat population, but they must have an effective com-

munications network; I've observed no casualties. Their movements are quick; they're among the competent thieves that steal choice tidbits from other birds at feeders.

The varied and melodic song of the goldfinch elicits widespread praise. It's difficult, indeed, to uncover any criticism of the bird. In the winter, the coloring of the male becomes much paler, but that's hardly a cause for fault-finding. During the colder months they're apt to be found in the company of other small finches such as redpolls and pine siskins, which they somewhat resemble.

The American goldfinch defends a territory the size of the

Size: 4½"L
Breeding range: Newfoundland west to British Columbia, south to southern California through Colorado to North Carolina

Winter range: Breeding range except northern portion to northern Mexico, Gulf states, and Florida
Preferred habitat: Fields, pastures, swamps
Preferred nest site: Leafy bushes or trees

Clutch size: 4 to 6, usually 5
Incubation period: 12 to 14 days
Nestling period: 11 to 15 days
Broods per season: 1 or 2
Preferred food: Insects, buds, seeds

evening grosbeak's by chases and circular flights. Later in the season, be on the lookout for an American goldfinch chasing a female. Sometimes the pursuit lasts for as long as twenty minutes, in a flat flight.

American goldfinches choose leafy bushes or trees, preferably with four upright branches or a fork, 1 to 33 feet from the ground, for their tightly woven nests of milkweed with rims of bark, the whole thing bound with spider silk. Sometimes they use materials from other nests. The female spends four to five days building the cup-shaped nest 2⅞ inches in diameter and 2¾ inches high.

American goldfinches feed their young regurgitated seeds, the male giving the female food for the nestlings. The nestling period is eleven to fifteen days. During the fledgling period (lasting up to a month) the male feeds the young directly. The female is usually nesting again.

FINCHES *Fringillidae*

Red Crossbill

(Loxia curvirostra)

WITH its unusual crossed mandibles, brick red plumage, and friendly nature, the red crossbill delights amateur and experienced birder alike. Always associated with cone-bearing trees, it prefers the seeds of white and pitch pines, native spruces, balsam, hemlock, and larch, but will also eat beech, elm, and maple seeds as well as sunflower and dandelion seeds. The red crossbill uses its feet parrot-style to climb trees and to hold cones, inserting its bill under

the scales and extracting the seeds with its tongue.

It builds its saucer-shaped nest near the ground in a conifer forest, constructing it out of twigs and roots and lining it with moss, hair, fur, and feathers. The female incubates the three to five eggs, which are greenish-white, spotted on the wide end with brown and lavender. If cones are abundant in the red crossbill's breeding range it may not migrate at all.

Size: 5½"L
Breeding range: Newfoundland to northeastern Pennsylvania west to Alaska, south through the western mountains to Central America

Winter range: Permanent resident in breeding range; wanders south to the Gulf coast
Preferred habitat: Coniferous forests
Preferred nest site: Conifer
Clutch size: 3 to 5, usually 4

Incubation period: 12 to 14 days
Nestling period: 15 to 17 days
Broods per season: 1
Preferred food: Seeds, especially of conifers, buds, wild fruits

APPENDIX

WHO EATS WHAT

SUNFLOWER SEEDS | Cardinal, goldfinch, purple finch, Harris's sparrow, white-crowned sparrow, dark-eyed junco, crossbill, cowbird, towhee, blackbird, grackle, mountain chickadee, black-capped chickadee, titmouse, yellow-rumped warbler, white-breasted nuthatch, red-breasted nuthatch, brown-headed nuthatch, grosbeak, jay.

PEANUT BUTTER | Cardinal, song sparrow, chipping sparrow, tree sparrow, field sparrow, dickcissel, mockingbird, black-capped chickadee, brown-capped chickadee, titmouse, yellow-rumped warbler, pine warbler, wood thrush, summer tanager, Carolina wren, red-bellied woodpecker, bluebird.

NUT MEATS, PEANUTS | Goldfinch, purple finch, house finch, white-throated sparrow, dark-eyed junco, bunting, towhee, red-winged blackbird,grackle, northern oriole, catbird, thrasher, black-capped chickadee, Carolina chickadee, mountain chickadee, titmouse, myrtle warbler, orange-crowned warbler, pine warbler, white-breasted nuthatch, hermit thrush, Carolina wren, red-bellied woodpecker, red-breasted nuthatch, brown-headed nuthatch, starling, kinglet, bluebird, yellow-bellied sapsucker, red-headed woodpecker, mourning dove, blue jay, Steller's jay, mockingbird.

MILLET | Purple finch, house finch, song sparrow, slate-colored junco, dickcissel, cowbird, red-winged blackbird, field sparrow, goldfinch, fox sparrow, mourning dove.

HEMP | Red-winged blackbird, tree sparrow, Harris's sparrow, goldfinch, cowbird, house finch, purple finch, song sparrow.

FRUIT | House finch, towhee, red-winged blackbird, summer tanager, yellow-bellied sapsucker, northern oriole, robin, western tanager, catbird, mockingbird, starling, red-bellied woodpecker, red-headed woodpecker, dove, bluebird, fox sparrow, Steller's jay, hermit thrush.

THISTLE | Finch, sparrow, chickadee, titmouse, towhee, junco, mourning dove.

CRACKED CORN | Red-winged blackbird, grackle, thrasher, blue jay, red-bellied woodpecker, red-headed woodpecker, house finch, song sparrow, tree sparrow, fox sparrow, field sparrow, Harris's sparrow, white-crowned sparrow, white-throated sparrow, dark-eyed junco, bunting, dickcissel, towhee, mourning dove, American crow.

SUET | Field sparrow, dark-eyed junco, house finch, song sparrow, chipping sparrow, tree sparrow, grackle, mockingbird, catbird, California thrasher, black-capped chickadee, mountain chickadee, brown-capped chickadee, orange-crowned warbler, yellow-rumped warbler, pine warbler, white-breasted nuthatch, red-breasted nuthatch, starling, kinglet, wood thrush, hermit thrush, summer tanager, Carolina wren, red-bellied woodpecker, red-headed woodpecker, yellow-bellied sapsucker, northern oriole, blue jay, titmouse.

WHEAT | Dark-eyed junco, dickcissel, Harris's sparrow, grackle.

OATS | Dickcissel, black-capped chickadee, chipping sparrow.

PUMPKIN SEEDS	Tree sparrow, purple finch, black-capped chickadee.
SQUASH SEEDS	Purple finch, black-capped chickadee, bluebird.
RAISINS	Wood thrush, bluebird, catbird, thrasher, mockingbird, red-winged blackbird, robin.
CHEESE	Catbird, thrasher, Carolina wren.
SCRATCH FOOD	Purple finch, cardinal, cowbird.
WEED SEEDS	White-throated sparrow, fox sparrow, dove, field sparrow.
MIXED SEEDS	Buntings, towhee, starling, kinglet, California thrasher, black-capped chickadee, red-bellied woodpecker.
GRAINS	Rusty blackbird, mockingbird, California thrasher, red-headed woodpecker, dove.
CHERRY & AUTUMN ELAEAGNUS	Cardinal, bluebird, flicker, catbird.
DATE PALMS	Bluebird, robin, waxwing, mockingbird.
MOUNTAIN ASH	Bluebird, flicker, waxwing, western tanager, hermit thrush.
BLACK GUM	Bluebird, robin, flicker, hermit thrush.
CAMPHOR TREES	Bluebird, robin.
CHINESE & SIBERIAN ELMS	Goldfinch, waxwing.
DOGWOOD & SUMAC	Hermit thrush, wood thrush, robin.
VIRGINIA CREEPER	Hermit thrush, flicker, robin, bluebird.
BITTERSWEET	Hermit thrush, bluebird, robin.
COTONEASTER	Robin, bluebird, waxwing.
BIRCH & ALDER	Fox sparrow, goldfinch.
CORNMEAL	Yellow-rumped warbler, pine warbler, wood thrush.
SUMAC & GOLDENROD SEEDS	Yellow-rumped warbler, wood warbler, robin.
SPRUCE, PINE, FIR, MAPLE SEEDS	White-breasted nuthatch, red-breasted nuthatch, red crossbill.
BERRIES	Indigo bunting, white-throated sparrow, towhee, chipping sparrow, starling, kinglet, bluebird, flicker, Carolina wren, northern oriole, California thrasher, mountain chickadee, cardinal, waxwing, wood thrush, mockingbird, black-capped chickadee, yellow-rumped warbler, red-headed woodpecker, robin, blue jay.

HABITAT REQUIREMENTS

BIRD	FOOD	NESTING MATERIAL	REMARKS
Allen's hummingbird	Nectar, small insects	Moss, plant stems and down, lichens, bark shreds, spiderweb	Nests in willows or cottonwoods in arid areas
American goldfinch	Insects, buds, seeds	Vegetable fibers, thistle down	Nests in leafy bushes or trees
American kestrel	Insects, small mammals, reptiles, amphibians, birds	None	Cavity nester; accepts birdboxes
American redstart	Insects	Bark shreds, leaf stalks	Nests in woodlands, swamps, gardens
American woodcock	Earthworms, larvae of insects	Dead leaves, twigs, pine needles	Nests on ground, not concealed
Anna's hummingbird	Small insects, nectar	Lichens, moss, feathers, fur	Often builds on branch near or over water
Bald eagle	Fish, small to medium-size mammals	Sticks, branches, grass, moss, sod, weeds	Enormous aerie, usually built in fork of giant tree
Barn owl	Mice	None	Accepts birdboxes
Barn swallow	Insects	Mud, straw	Nests in buildings, bridges, culverts; formerly on cliffs, in caves
Belted kingfisher	Fish, crayfish, insects	None	Excavates burrow in bank, near water
Bewick's wren	Insects	Bark, assorted rubbish	Crevice nester; accepts birdboxes
Black vulture	Carrion	None	Ground nester; both parents incubate
Black-and-white warbler	Insects	Grass, rootlets	Ground nester, at base of tree, stump, or rock
Black-backed woodpecker	Insects	Wood chips	Cavity nester, tree or pole
Black-billed cuckoo	Insects, fleshy fruits	Twigs	Nests in second growth in shrubs
Black-billed magpie	Omnivorous — insects, vegetable matter, carrion	Sticks, grass	Colonial nester; prefers willow thickets
Black-capped chickadee	Insects, seeds, fruits	Moss, vegetable fibers, cinnamon fern, wool, feathers, hair, fur	Accepts birdboxes
Black-chinned hummingbird	Nectar, small insects	Bud scales, leaves, plant down, spiderwebs	Nest looks like a little yellow sponge
Black-headed grosbeak	Insects, fruits	Twigs, plant stems	Nests in bush or tree
Blue grosbeak	Insects, grains	Grass, rootlets, snakeskin	Nests in shrub or on low branch
Blue jay	Omnivorous — seeds, fruits, acorns, young mice, nestlings	Twigs, rootlets, paper, rags, string, bark, grass, leaves, feathers	Careless cup-shaped nest in crotch of tree
Blue-gray gnatcatcher	Insects	Plant down, spiderweb, lichens	Nests in tall tree in open woods, gardens
Bobolink	Insects, seeds	Grass, weed stems	Ground nester in tall grass
Bobwhite	Vegetation, seeds, insects	Weeds, grass	Nests in hollow in grass
Boreal chickadee	Insects, seeds, berries	Moss, lichens, bark, fern down, feathers, fur	Cavity nester
Brewer's blackbird	Insects, seeds	Twigs, bark, mud	Often nests in colonies
Brown creeper	Insects, insect and spider eggs and larvae	Twigs, leaves, grass, bark, moss	Nests under loose flake of bark on tree

Bird	Food	Nesting Material	Remarks
Brown thrasher	Insects, berries, fruits, grains	Twigs, dry leaves, grass	Ground nester; new mate for second brood
Brown-headed cowbird	Insects, seeds, berries, grains	None	Parasite
Brown-headed nuthatch	Insects, pine seeds	Bark shreds, fur, feathers, pine-seed wings, grass, cotton, wool, pine needles	Cavity nester in pine woods, cypress swamps
Bushtit	Insects and larvae	Moss, lichens, leaves, spiderwebs	Will nest in city gardens in bush or tree
California quail	Insects, grains, berries, fruits	Grass	Nests in slight hollow under brush heaps, beside a rock, sometimes in gardens
California thrasher	Wild fruits, berries, insects	Coarse twigs	Nests near ground; nest similar to mockingbird's but larger
Canada goose	Grasses, marsh plants, aquatic plants, grains	Sticks, reeds, grass, down	Nests on ground near water
Carolina chickadee	Insects, seeds, fruits	Moss, grass, bark, plant down, feathers, hair, fur	Parent pulls down over eggs when absent from nest; accepts birdboxes
Carolina wren	Insects	Twigs, grass, leaves	Accepts birdboxes, 10 feet or less from ground
Cattle egret	Insects, reptiles, amphibians, crustaceans	Sticks	Nests in bushes or trees, not necessarily near water; singly or in small colonies
Cedar waxwing	Berries, insects	Grass, twigs, weeds, string, yarn, rootlets	Prefers open woods
Chimney swift	Flying insects	Twigs	Will repair and use old nests
Chipping sparrow	Insects, seeds	Twigs, weeds, grass, rootlets, hair	Nests in vines or shrubs
Clark's nutcracker	Omnivorous — piñon nuts	Twigs, grass	Nests in conifers
Cliff swallow	Flying insects	Mud	On outside of buildings, on cliffs, bridges
Common grackle	Insects, fruits, grains	Twigs, coarse grass or seaweed, mud, soft grass	Somewhat colonial; sometimes uses osprey nest
Common ground dove	Seeds, grains, berries	Twigs, pine needles, rootlets, grass	Nests on ground or in tree
Common loon	Fish, amphibians, insects, aquatic plants	Rushes, grass, twigs, reeds	Likes small wooded islands, privacy
Common nighthawk	Flying insects	None	Nests on bare ground, flat roofs
Common yellowthroat	Insects	Grass, leaves, hair	Likes a moist, shrubby area
Crow	Omnivorous — grains, insects, carrion	Sticks, bark, vegetable fibers, rootlets, grass, feathers, seaweed, cornstalks	Great horned owls sometimes take crows' nests; long-eared owls may use abandoned one
Dark-eyed junco	Insects, wild fruits, seeds	Grass, rootlets, moss, bark shreds, cow or deer hair	Bulky nest on ground; builds new one for second brood
Dickcissel	Seeds, grains, insects	Weeds, grass	Builds large, bulky nest
Downy woodpecker	Insects	Wood chips	Accepts birdboxes

BIRD	FOOD	NESTING MATERIAL	REMARKS
Eastern bluebird	Insects, fruits	Fine grass	Accepts birdboxes
Eastern kingbird	Flying insects, wild fruits	Weeds, moss, bark, feathers, cloth, string	Nest is poorly constructed, easily destroyed
Eastern meadowlark	Insects, seeds, grains	Grass, weeds	Ground nester in pastures, fields, marshes
Eastern phoebe	Flying insects	Weeds, grass, mud, moss, hair	Accepts nesting platforms; builds on shelflike projections
Eastern wood-pewee	Insects	Plant fibers, lichens	Builds shallow nest on horizontal branch
European starling	Insects, seeds, fruits, grains	Mostly grass	Will commandeer birdboxes intended for other species
Evening grosbeak	Buds, fruits, seeds, insects	Twigs, grass, rootlets	Relatively social nester; nests in conifers
Field sparrow	Insects, seeds	Grass, seeds, rootlets	Ground nester
Fox sparrow	Insects, seeds, fruits	Grass, moss, leaves, bark, weed rootlets, feathers, rootlets	May nest on the ground or up to 12 feet high in bush or tree
Gambel's quail	Insects, grains, berries, fruits	Grass, feathers	Nests in scrape at base of shrub
Gila woodpecker	Flying insects, ants, berries, corn	Wood chips	Nests in cavity in saguaro cactus or cottonwood
Golden eagle	Small to medium-size mammals, fish, reptiles, carrion	Sticks, grass	Nests in high tree or on cliff
Golden-crowned kinglet	Insects	Moss, lichens	Nest is suspended from twig of conifer
Golden-crowned sparrow	Seeds	Twigs, moss, grass, fine roots	Nests in depression in ground or on tussock
Gray catbird	Insects, fruits, berries	Twigs, weeds, leaves, grass, rootlets	Likes to nest in briars, may use same nest for second and third broods
Gray jay	Omnivorous — insects, fruits, seeds, buds	Twigs, grass, bark, moss, lichens, plant down, feathers, cocoons, spiderweb	Nests early, late winter, early spring
Great blue heron	Insects, fish, amphibians	Sticks	Colonial; nests in swamp trees
Great horned owl	Small mammals, birds, reptiles	None	Often uses deserted hawk or crow's nest, or lays eggs on ground amidst old bones, skulls, fur, etc.; deep woods
Green-backed heron	Small fish, crustaceans, insects	Sticks	Colonial; tree nester
Green-tailed towhee	Seeds, wild fruits, insects	Grass, bark shreds	Nests in underbrush
Hairy woodpecker	Adult and larval beetles, ants, fruits, nuts, corn	Wood chips	Accepts birdboxes
Hermit thrush	Insects, fruits	Twigs, grass, bark, weeds, ferns, moss, pine needles, rootlets, porcupine hair	Ground nester in damp, cool woods
Herring gull	Carrion, garbage, marine animals	Grass, seaweed, shells, feathers	Nests in sandy area, on ground
Horned lark	Insects, seeds	Grass	Ground nester
House finch	Seeds, fruits, insects	Twigs, grass, debris	Will nest anywhere

BIRD	FOOD	NESTING MATERIAL	REMARKS
House sparrow	Insects, seeds, garbage	Grass, straw, weeds, assorted junk	Will commandeer birdboxes intended for other species
House wren	Insects	Twigs, grass, rootlets, feathers, hair, rubbish	Accepts birdboxes
Indigo bunting	Insects, seeds	Twigs, grass, weeds, leaves, hair, feathers	May use same nest for five years, repairing as necessary
Killdeer	Insects	None	Nests in open, on ground or on roofs, on pebbles, wood chips, grass
Lazuli bunting	Seeds, insects	Grass, leaves	Nests on low branch near water
Least flycatcher	Flying insects	Plant fibers, grass	Nests in upright fork of tree
Lewis's woodpecker	Sap, insects	Wood chips	Cavity nester
Loggerhead shrike	Insects, small animals, and birds	Twigs, grass	Nests in dense brush
Magnolia warbler	Insects	Twigs, grass	Prefers small conifer in woods edges, gardens
Mallard	Seeds, leaves	Reeds, grass, feathers, down	Likes to nest on dry ground in tall grass or alfalfa
Mountain bluebird	Insects, fruits	Grass	Accepts birdboxes
Mountain chickadee	Insects, seeds, berries	Grass, moss, plant down, bark, rootlets, fur, hair, wool	Cavity nester in high evergreen forests
Mourning dove	Seeds, grains	Twigs	Nests on ground or in tree
Mute swan	Crustaceans, insects	Sticks, roots, trash	Nests on small islet in shallow, marshy margin of pond or on bank near water
Northern cardinal	Seeds, fruits, grains, insects	Twigs, leaves, weeds, grass, hair, rootlets	Usually nests within 8 feet of ground
Northern flicker	Ants, other insects, wild fruits	Wood chips	Accepts birdboxes
Northern harrier	Small mammals and birds	Grass, reeds	Ground nester
Northern mockingbird	Insects, fruits, berries, seeds	Twigs, leaves, moss, hair, rootlets	Nests in shrubs or trees
Northern oriole	Insects, fruits	Weed stalks, inner bark, string, plant down, hair	Gourd-shaped nest, 5 inches long
Orange-crowned warbler	Insects	Grass, rootlets	Nests on ground in shrubs on hillside
Orchard oriole	Insects, fruits	Weeds, fibers, dry grass	Likes a hardwood tree; tendency to colonial nesting
Osprey	Fish	Sticks, bark, sod, grass	Loose colonies; likes trees and poles, high or low
Painted bunting	Seeds, insects	Twigs, leaves, rootlets, bark, weeds, grass, caterpillar silk, horsehair	Likes to nest in thick foliage near ground
Palm warbler	Insects, berries	Grass	Nests in moss at foot of tree or shrub, in bogs or on lawns
Peregrine falcon	Small to large birds	None	Nests in scrape on ground or appropriates nest of buzzard, raven, eagle

BIRD	FOOD	NESTING MATERIAL	REMARKS
Pileated woodpecker	Larvae and adults of carpenter ants and other insects, wild fruits, acorns, beechnuts	Wood chips	Excavates new hole for each brood
Pine siskin	Insects, buds, seeds	Twigs, bark, hair, plant down, moss, fur	Conceals nest in evergreen
Plain titmouse	Insects, acorns, berries	Cotton, wool, feathers; anything soft	Accepts birdboxes
Purple finch	Seeds, buds, fruits, insects	Twigs, weeds, grass	Prefers evergreens
Purple martin	Flying insects	Grass, twigs, bark, paper, leaves, string	Nests in dense colonies; accepts martin houses
Pygmy nuthatch	Insects	Feathers, plant down, wool, fur	Nests behind bark crevices or in cavities, preferably in conifers
Red crossbill	Seeds, especially of conifers, buds, wild fruits	Evergreen twigs, moss	Usually nests in conifer
Red-bellied woodpecker	Insects, mast, corn, wild fruits	Wood chips	Excavated nest is gourd shaped
Red-breasted nuthatch	Insects, seeds	Wood chips, feathers, grass, rootlets, bark	Accepts birdboxes
Red-eyed vireo	Insects	Bark, grass, spider egg cases, plant down	Nests on branch 5 to 15 feet high
Red-headed woodpecker	Insects, acorns, wild fruits	Wood chips	Both parents excavate cavity, in dead or live tree; accepts birdboxes
Red-shafted flicker	Ants, berries, acorns	Wood chips	Accepts birdboxes
Red-tailed hawk	Small mammals, amphibians, reptiles, nestlings, insects	Twigs	Builds on rock ledges, in trees, in open situation; may use same nest repeatedly
Red-winged blackbird	Insects, seeds, grains	Rushes, grass, moss, milkweed fiber	Social nester, in marshy areas; will strike at intruders
Ring-necked pheasant	Grains, seeds, vegetation	Weeds, grass, leaves	Ground nester
Roadrunner	Reptiles, rodents, insects	Twigs	Nests in cactus, mesquite
Robin	Fruits, earthworms, insects	Twigs, mud, fine grass, animal fur	Will accept nesting platform; will nest on and around house in evergreen and deciduous trees
Rock dove	Seeds, grains, handouts	Twigs, grass, straw, debris	Male gathers materials, female builds
Rose-breasted grosbeak	Insects, seeds, fruits	Twigs, weeds, grass	Prefers fork of deciduous tree as nest site
Ruby-crowned kinglet	Insects, seeds, fruits	Moss, lichens	Nest on, or suspended from, limb of conifer
Ruby-throated hummingbird	Nectar, small insects, sap	Lichens, bud scales, plant down, spider silk	Nest the size of eyecup
Ruffed grouse	Seeds, insects, fruits	Leaves, pine needles	Nests in a hollow under log or at base of tree, in woods
Rufous hummingbird	Nectar, insects	Plant down, vegetable stalks, moss, bark shreds, lichens	Sometimes builds new nest on top of old one
Rufous-sided towhee	Insects, seeds, fruits, mast	Leaves, weeds, bark, grass, twigs, ferns, moss, pine needles, grass rootlets	Nests on or near ground

BIRD	FOOD	NESTING MATERIAL	REMARKS
Sage thrasher	Insects, berries, fruit	Bark strips, twigs, grass	Nests in sagebrush
Saw-whet owl	Small mammals, insects	None	Uses woodpecker or squirrel holes; accepts birdboxes
Say's phoebe	Flying insects	Mud, moss, wood	Nests on ledges, bridges, buildings
Scarlet tanager	Insects, fruits	Twigs, rootlets, grass	Small, flimsy, flat cup-shaped nest, preferably on oak limb
Scott's oriole	Nectar, insects, fruits	Grass, yucca threads, horsehair, cotton waste, grass	Prefers to nest in yucca, near water
Screech owl	Small rodents and insects	None	Accepts birdboxes; lays eggs on leaves or any rubble in cavity
Scrub jay	Omnivorous — insects, acorns, young birds	Twigs, rootlets	Loose colonies of up to six nests
Sharp-tailed grouse	Vegetation, insects	Grass, feathers	Nests in hollow in ground
Song sparrow	Insects, seeds, fruits	Grass, bark, leaves, hair, rootlets	Likes shrubs, high weeds as nest site
Spotted sandpiper	Insects	Grass, leaves, weed stems	Saucer-shaped nest on ground, near water
Spruce grouse	Buds and needles of conifers, berries, insects	Dry twigs, leaves, moss, grass	Ground nester; prefers site under low conifer branch
Steller's jay	Omnivorous — acorns, fruits, seeds, berries	Sticks, mud, grass, rootlets, pine needles	Likes to use platform of old leaves in crotch of tree
Summer tanager	Insects	Grass, rootlets, leaves, bark, Spanish moss	Shade trees, woodlands
Swainson's thrush	Insects, wild fruits	Grass, moss, twigs	Nests in spruce
Three-toed woodpecker	Insects	Wood chips	Nests in cavity in dead conifer
Townsend's solitaire	Flying insects	Grass, pine needles	Nest well concealed on ground
Tree sparrow	Seeds	Grass, rootlets, weeds, feathers, hair, fur	Nests in thickets
Tree swallow	Flying insects, berries, seeds	Grass, feathers	Accepts birdboxes
Tufted titmouse	Insects, seeds, mast, fruits	Leaves, bark, moss, grass, cotton, wool, feathers, hair, fur, old snakeskins	Accepts birdboxes
Turkey	Acorns, berries, plants, grains, insects	Dead leaves	Ground nester
Turkey vulture	Carrion	None	Ground nester, preferably in gravel or sawdust
Veery	Insects, wild fruits, seeds	Grass, twigs	Ground nester
Violet-green swallow	Flying insects	Dry grass, feathers	Accepts birdboxes
Warbling vireo	Insects	Moss, bark, grass	Nests in tall trees in gardens, woods edges
Western bluebird	Insects, fruits, berries, weed seeds	Grass	Accepts birdboxes
Western kingbird	Flying insects	Twigs, grass	Often nests on human-built structure
Western meadowlark	Insects, grains	Grass, weeds	Ground nester in pastures, fields, marshes
Western tanager	Insects, fruits, berries	Grass, bark, twigs, hair, rootlets	Fragile nest, saucer shaped

Bird	Food	Nesting Material	Remarks
Whip-poor-will	Flying insects	None	Lays eggs on dead leaves on ground
White-breasted nuthatch	Insects, seeds, fruits, mast	Bark, grass, twigs, rootlets, feathers, hair	Accepts birdboxes
White-crowned sparrow	Insects, seeds	Twigs, grass, leaves, bark, rootlets, moss, fur, hair	Ground nester
White-eyed vireo	Insects, wild fruits	Leaves, moss, wasp paper, sticks, soft woody fibers	Nests in shrub, often near water
White-throated sparrow	Insects, seeds, wild fruits	Grass, twigs, leaves, moss, rootlets, hair	Ground nester
White-throated swift	Flying insects	Twigs	Nests on dry cliffs
Willow flycatcher	Flying insects	Shreds of plant materials, grass	Prefers crotch of shrub, near ground
Winter wren	Insects	Sticks, moss	Nests in tangled growth near ground
Wood duck	Acorns, insects	Wood chips, down	Cavity nester; accepts birdboxes
Wood thrush	Insects, fruits	Grass, leaves, weeds, moss, mud	Likes something white on outer nest wall
Yellow warbler	Insects	Plant fibers	Nests in fork of sapling or shrub in thickets, gardens, farmlands
Yellow-bellied sapsucker	Sap, insects, fruits, berries	Wood chips	Cavity nester
Yellow-billed cuckoo	Insects, wild fruits	Twigs	Nests 4 to 10 feet from ground; likes thickets
Yellow-breasted chat	Insects, berries	Grass, leaves	Nests near ground, in thicket
Yellow-headed blackbird	Insects, grains, seeds	Sedge, grass	Attaches nest to reeds near ground in marsh
Yellow-rumped warbler	Insects, berries	Dried grass, weeds, bark, spiderweb, hair, plant down	Nests in spruce or fir forests or gardens
Yellow-throated vireo	Insects	Spider silk, lichens, moss, spider egg cases	Nests in shade trees, orchards
Yellowthroat	Insects	Twigs, bark strips	Nests in pine, oak, sycamore, in bottom-lands

ALL ABOUT EGGS

Notes are on page 301.

BIRD	BASE COLOR	DECORATION
American goldfinch	Bluish white	None
American woodcock	Pinkish brown	Light-brown spots or blotches overlaid with darker brown markings
Bald eagle	Dull white	None
Barn owl	White	None
Barn swallow	White	Spotted and dotted with brown
Belted kingfisher	White	None
Bewick's wren	White	Fine reddish brown spots
Black vulture	Gray-green, bluish white, or dull white	Large splotches or spots of pale chocolate or lavender wreathed or clustered at the large end and overlaid with dark brown blotches and spots
Black-capped chickadee & Carolina chickadee	White	Spotted and dotted with reddish brown concentrated at larger end
Blue jay	Olive or buff	Dark brown or grayish dots, spots, and blotches
Bobwhite	Dull or creamy white	None
Brown creeper	White or creamy	Finely dotted with reddish brown; sometimes wreathed
Brown thrasher	Whitish with a blue or green tinge	Heavily and evenly spotted or dotted with reddish brown
Brown-headed cowbird	White or grayish	Evenly dotted with browns, more heavily at the larger end
California thrasher	Light greenish blue	Flecked with brown
Canada goose	Creamy or dirty-looking white, dull yellowish green	None
Carolina wren	Creamy or pinkish	Large reddish brown spots
Cedar waxwing	Pale gray	Lightly and irregularly spotted with brown and blotched with brownish gray
Chimney swift	White[7]	None
Chipping sparrow	Pale bluish green	Dotted, spotted, blotched, and scrawled with dark brown, black, and purple, especially at larger end
Clark's nutcracker	Pale greenish	Tiny dots of brown or olive
Common grackle	Pale greenish white or pale yellow-brown	Blotched, streaked, and spotted with dark browns and purple
Common loon	Greenish or brownish	Scattered spots or blotches of brown or black
Common raven	Greenish	Brown or olive markings in various patterns
Crow	Bluish or grayish green	Irregularly blotched and spotted with brown and gray
Dark-eyed junco	Pale bluish white or grayish	Dotted, spotted, and occasionally blotched with browns, purple, and gray, especially at larger end[23]
Downy woodpecker	White	None

BIRD	BASE COLOR	DECORATION
Eastern bluebird	Bluish, bluish white, or pure white	None
Eastern kingbird	Creamy white	Heavily and irregularly spotted with brown, black, and lavender; sometimes wreathed
Eastern phoebe	White	One or two may be sparsely spotted
European starling	Pale bluish or greenish white	None
Gray catbird	Greenish blue	None
Ground dove	White	None
Hairy woodpecker	White	None
House finch	Pale bluish green	Sparingly spotted and dotted with black
House sparrow	White or greenish white	Dotted and spotted with gray
House wren	White	Thickly speckled with tiny reddish or cinnamon dots, especially at larger end
Indigo bunting[20]	White or pale bluish white	None
Killdeer	Buff	Bold black or brown spots, scrawls, and blotches; sometimes wreathed or capped
Mallard	Light greenish or grayish buff to white	None
Mourning dove	White	None
Northern cardinal	Grayish blue, or greenish white	Dotted, spotted, and blotched with browns, grays, and purples, sometimes heavily
Northern flicker	White	None
Northern mockingbird	Bluish green	Heavily marked with reddish brown spots and blotches; sometimes wreathed; sometimes paler or darker, perhaps a buffy gray, with yellow, gray, chocolate, and purple spots
Northern oriole	Pale bluish or grayish white	Irregularly streaked, scrawled, and blotched with browns, lavender, and black, especially at the larger end
Orchard oriole	Bluish white	Heavily blotched with brown, purple, and lavender at the larger end
Osprey	White or pale pink	Heavily spotted or blotched with a rich or reddish brown
Pileated woodpecker	White	None
Pine siskin	Pale bluish green	Spotted and speckled at the larger end with purple or black
Purple martin	White	None
Red-breasted nuthatch	White	Reddish brown spots — less heavily marked than eggs of white-breasted nuthatch
Red-headed wood-pecker	White	None

Bird	Base color	Decoration
Red-winged blackbird	Pale bluish green	Spotted, blotched, marbled, and scrawled with browns, purples, and black, especially at the larger end
Ring-necked pheasant	Rich, brownish olive or olive-buff	None
Robin	Blue	None
Rock dove	White	None
Rose-breasted grosbeak	Bluish green or grayish	Spotted and blotched with lilac and brown[21]
Ruby-throated hummingbird	White	None
Ruffed grouse	Buff	Some are speckled with brownish spots
Rufous-sided towhee	Creamy grayish, or pinkish white	Finely and evenly dotted and spotted reddish brown; wreathed or capped at larger end
Scarlet tanager[20]	Pale blue or green	Irregularly dotted, spotted, and blotched with brown, especially at larger end; sometimes capped
Screech owl	White	None
Scrub jay (Florida and Rocky Mountains)	Greenish	Blotched and spotted with irregular brown or cinnamon markings; wreathed
Song sparrow	Greenish white	Heavily dotted, spotted, and blotched with reddish brown and purple, sometimes underlaid with gray[23]
Steller's jay	Pale greenish blue	Spotted with browns, purple, and olive
Tree swallow	White	None
Tufted titmouse	White or creamy	Evenly spaced fine brown dots, especially at the larger end
Turkey	Pale buff or buffy white	Evenly marked with reddish brown or pinkish buff spots or fine dots
Turkey vulture	Dull or creamy white	Irregular spots, blotches, and splashes of pale brown with overlay of bright brown
Whip-poor-will	White	Irregularly spotted and blotched with gray and overlaid with brown
White-breasted nuthatch	Creamy white	Some are heavily marked, especially at the larger end, with reddish brown dots
White-crowned sparrow	Greenish or bluish white	Heavily spotted with reddish or purplish brown, with suggestion of a wreath
White-throated sparrow	Whitish, bluish, or green	Spotted with reddish brown
Wood duck	Creamy, dull white, or pale buff	None
Yellow-rumped warbler	White	Speckled brown and purple

WHO INCUBATES THE EGGS

American goldfinch	Female[12]
American woodcock	Female[4]
Bald eagle	Both parents
Barn owl	Female[6]
Barn swallow	Both parents
Belted kingfisher	Both parents
Bewick's wren	Mostly female[12]
Black vulture	Both parents
Black-capped chickadee	Female[10]
Blue jay	Both parents
Bobwhite	Both parents
Brown creeper	Both parents
Brown thrasher	Both parents[14]
Brown-headed cowbird	Host bird[19]
California thrasher	Both parents
Canada goose	Female[10]
Carolina chickadee	Female
Carolina wren	Mostly female
Cedar waxwing	Female
Chimney swift	Both parents, sometimes simultaneously
Chipping sparrow	Usually female
Common grackle	Both parents in the East and South; female elsewhere
Common loon	Mostly female
Common raven	Female
Crow	Both parents
Dark-eyed junco	Female or both parents (authorities disagree)
Downy woodpecker	Both parents
Eastern bluebird	Female
Eastern kingbird	Female
Eastern phoebe	Female
European starling	Both parents
Gray catbird	Female[15]
Ground dove	Both parents
Hairy woodpecker	Both parents
House finch	Female
House sparrow	Female[17]
House wren	Female[13]
Hummingbirds (North American)	Female
Indigo bunting[20]	Mostly female
Killdeer	Both parents

Mallard	Female
Mourning dove	Both parents
Northern cardinal	Female
Northern flicker	Male at night and both parents take turns during the day[9]
Northern mockingbird	Female
Northern oriole	Female
Orchard oriole	Female[18]
Osprey	Female
Pileated woodpecker	Both parents
Purple martin	Female
Red-breasted nuthatch	Mostly female[12]
Red-headed woodpecker	Both parents
Red-winged blackbird	Female
Ring-necked pheasant	Female
Robin	Female[16]
Rock dove	Both parents
Rose-breasted grosbeak	Both parents
Ruby-throated hummingbird	Female
Ruffed grouse	Female
Rufous-sided towhee	Female[22]
Scarlet tanager[20]	Female
Screech owl	Mostly or exclusively female
Scrub jay (Florida and Rocky Mountains)	Female
Song sparrow	Female
Steller's jay	Both parents
Tree swallow	Female
Tufted titmouse	Female or both parents (conflicting sources)
Turkey	Female[3]
Turkey vulture	Both parents
Whip-poor-will	Female
White-breasted nuthatch	Female[13]
White-crowned sparrow	Female
White-throated sparrow	Female
Wood duck	Female
Yellow-rumped warbler	Female

Key to Egg charts

1. 3½ × 2½ inches.
2. Clutches can contain twenty eggs. One nest reported with thirty-seven, but perhaps occupied by two females.
3. Up to twenty eggs have been found in a nest, but may indicate more than one female.
4. Female may abandon nest if disturbed.
5. Two eggs, laid thirty-six hours apart.
6. Female begins incubation at once; eggs hatch at different times, from twenty-one to thirty-four days.
7. Black swift, Vaux's swift, and white-throated swift all lay white eggs.
8. Two pea-size eggs, laid forty-eight hours apart.
9. Eggs are laid between 5 and 6 A.M.
10. If disturbed at the nest, female fluffs her wings, hissing and swaying like a snake, with her beak open.
11. Similar to chickadee eggs.
12. Male feeds female.
13. Incubation lasts twelve to fifteen days, usually thirteen days.
14. Female spends three times as many hours on the nest as the male; sometimes they raise a cowbird.
15. Eggs are greener and smaller than robin eggs; catbirds will eject a cowbird egg.
16. Robins will eject a cowbird egg.
17. Twelve- to thirteen-day incubation period; possibly only ten to twelve days.
18. Male feeds female; orchard orioles often nest in same tree with kingbirds.
19. May lay three or four clutches; incubation takes eleven to twelve days.
20. Frequently parasitized by cowbirds.
21. Eggs similar to scarlet tanager eggs, but bigger and more heavily marked; male feeds female when she is sitting.
22. If sufficiently disturbed, female will use trailing-wing tactic to divert attention from the nest.
23. Considerable variation in appearance of eggs.

For Further Reading

Books on Birds

Baldwin, Edward A. The Weekend Workshop Collection of Birdfeeders, Shelters, & Baths: Over 25 Complete Step-by-Step Projects for the Weekend Woodworker. Pownal, VT: Storey Publishing, 1990.

Ramuz, Mark, and Frank Delicata. Birdhouses: 20 Step-by-Step Woodworking Projects. Pownal, VT: Storey Publishing, 1996.

Rupp, Rebecca. Everything You Never Learned about Birds. Pownal, VT: Storey Publishing, 1995.

Books on Gardening

Abraham, Doc & Katy. Green Thumb Wisdom: Garden Myths Revealed! Pownal, VT: Storey Publishing, 1996.

Art, Henry W. A Garden of Wildflowers: 101 Native Species and How to Grow Them. Pownal, VT: Garden Way Publishing, 1986.

Binetti, Marianne. Tips for Carefree Landscapes: Over 500 Sure-Fire Ways to Beautify Your Yard and Garden. Pownal, VT: Garden Way Publishing, 1990.

Denckla, Tanya. The Organic Gardener's Home Reference: A Plant-by-Plant Guide to Growing Fresh, Healthy Food. Pownal, VT: Garden Way Publishing, 1994.

Editors of Garden Way Publishing. The Big Book of Gardening Skills. Pownal, VT: Garden Way Publishing, 1993.

Glattstein, Judy. Waterscaping: Plants and Ideas for Natural and Created Water Gardens. Pownal, VT: Garden Way Publishing, 1994.

Kennedy, Des. Nature's Outcasts: A New Look at Living Things We Love to Hate. Pownal, VT: Storey Publishing, 1993.

Powell, Eileen. From Seed to Bloom: How to Grow over 500 Annuals, Perennials, & Herbs. Pownal, VT: Garden Way Publishing, 1995.

Riotte, Louise. Sleeping with a Sunflower: A Treasury of Old-Time Gardening Lore. Pownal, VT: Garden Way Publishing, 1987.

Rupp, Rebecca. Red Oaks & Black Birches: The Science & Lore of Trees. Pownal, VT: Garden Way Publishing, 1990.

Index

Italic page numbers indicate illustrations.
Boldface page numbers indicate charts.